From Student to Engineer

A College Course and its Textbook that Gives Students an Overview for those Considering the Mission which Embarks upon the Journey of the Engineer

Written by:
Brian Shell

2

Syllabus

(Table of Textbook Contents)

Course: "From Student to Engineer"

Special Thanks and Gratitude

I'm grateful to Lawrence Technological University.

The students and faculty encouraged me to write a textbook.

They also asked me to consider creating a course.

This book is the result.

Thank you!

Additional praise goes out to Ari Weinzweig and John Turnes.

Their contextual additions are most appreciated.

Gratzi gentlemen!

Dedication

This book is dedicated in loving memory to Zaven Margosian.

When I went to undergrad, he was the Dean of Arts & Sciences.

At Lawrence Tech, he was tough… but fair.

He took pride in the littlest things.

An example of that is the fact that he'd made sure every single student got their actual diploma with their name in it when they crossed the stage. It was a Herculean task. Most schools just hand out the diploma cover and then mail out the actual diploma later. That's the kind of attention-to-detail that separated him from the rest… and made our school just a little bit better… a little more special than most.

In 1995, when I left my corporate engineering job, he heard about it and my reason… that I wanted to become a writer. He immediately asked, "To write a textbook?"

Now, 20 years later: *"Zaven, I dedicate this textbook to you."*

So long ago, you planted a seed.

It has now started to sprout.

Rest in peace Zaven Margosian.

May your fine memory live forever.

Opening Quote

"You're only worth what you negotiate."

Said by the late guitarist James Gurley...
...as related by drummer Muruga Booker.

Introduction

First, let me start by saying that this book will not be like any other textbook you'll probably ever encounter. That's because there's no college course that I'm aware of that teaches you what your life will be like... or might be like... if you do decide to embark on the journey to become an engineer. It is an endless river from the sea of knowledge.

It provides the perspectives no one tells you when you start in college... to see if it's really a path for you. In other words, it presents the "I wish I knew then what I know now" approach of **"should I embark on this mission if I so decide to accept it?"**

Also, I'm not aware of many textbooks that use first-person point-of-view as I plan on using. So be prepared. I plan on trying to make this class I'm designing an outside-of-the-box kind of experience that lets the genie out of the bottle when it comes to student perspectives, college realities, and real-life lessons-learned which I hope will enable you to make the best decision if engineering is really for you. In other words, by reading this book and taking this course, I want you to learn that "No matter how far down the wrong road you go, turn back."

However, if you're a math and science geek like I was when I entered into college, than this book may be for you... and right up your alley. In the end of the movie, *"Revenge of the Nerds,"* the nerds of the school win... simply because they think smarter, not harder.

That's what I want from you... to start seeing in new ways... to envision the invisible... to create the new invention which makes life easier (and amasses you a nice fortune along the way... because it's nice to have money become "one less thing").

Also, after being a straight-A student through undergrad and graduate school and then working for a large aerospace corporation for three years before departing for years to become a writer, a musician, and an artist, you'll probably notice that I've developed a very unique style to my writing. For example, I like using the three-dot ellipses (...) to place conscious pauses in my writing. I like going off on personal tangents that switch the style from third-person to first-person when I want to show an example that drives home the point I want to make with clarity. After all, a good parable or metaphor can make all the difference in you really and truly grasping an elusive concept. Also, the more you can communicate in ways that enable a ten-year-old to understand a difficult concept, the more you will succeed in becoming a success in life... because technical communication is notoriously an area in which engineers fall short. As I've experienced first-hand, engineers often like explaining their work with a "taking a drink from a fire hydrant" approach... which is one reason we often get stereotyped as "boring or dull" when it comes to our dating experiences later on in life. As I've been told a few times while out on a first date, "Babysteps." In other words, give them the appetizer... save the entrée. That's because we have a tendency of saying too much to show our brilliance.

In the real world, less is always more.
In engineering, more is often more.
What a conundrum to face, eh?

So if you decide to embark upon this course and its accompanying text, bear with me. I don't pretend to know it all. As you'll learn if you decide to go to grad school, "the more you learn, the more you'll learn how much there is to learn."

Yet, I will try to share as much as I can so you have a robust understanding of what you're really embarking upon if you do decide to get your engineering degree. In many cases, you can go through years of college, only to later learn that its "life" really isn't for you... (probably) much to the dismay of your parents... who perceive engineering as a "respectable" profession to enter into. However, that doesn't mean it may be for you... and there's nothing wrong with realizing that... hopefully, sooner than later.

On the other hand though, going through school to become an engineer is a lot easier earlier in life than trying to navigate its waters later in life.

There's a saying, "Once an engineer, always an engineer."
I believe that to be true. Honestly, I'm glad I became an engineer.

As many people have agreed with me, it's a lot easier for an engineer to turn into an artist or another creative type than it is for an artist to turn into an engineer.

There's something that happens when you immediately go from high school into college. You get trained in learning a certain way of scholastic thinking before the "real world" starts to train you with *their* ways of doing things. As you might encounter later in life, one of the worst mottos you might face out there in the corporate life (such as the cartoon "Dilbert" often lampoons) is, "Well that's the way we've always done it."

That kind of thinking stagnates the imagination. It can stifle you from becoming the next trend-setting visionary who creates the next new best thing which may enable you to become the next flavor-of-the-month media-darling who creates an empire from a simple idea and a new way of doing things that makes this world a better place to be.

The point to remember is that accountability for "correctness" is crucial in the working world. As a student you don't come to college to test yourself and turn in your own test scores. No. The point of the professor is to first "calibrate" that correctness into the right arena which you then "tune" as you choose the courses which most peak your interests. That enables you to find fulfillment… even in the menial tasks. It grooms maturity.

Thus, this textbook offers **an overview**. It's like running a race and studying the boundaries of "the race" before you travel a single step. You learn the straight-aways, the uphill climbs, and the bodies of water you need to cross in order to be prepared.

Week 1:

An Undergraduate Engineering Student's First Two Years

Some Brief Brainstorming:

Education is fundamental to society.

Commuter schools in the winter time aren't as much fun. On the other hand, living on campus and having to walk a long way in the snow isn't fun either.

Just realize, in life, it's always something.

In life, we don't dodge the storms. We learn to dance in the rain.

The Humor in this Stage:

When I was working as a clerk at a party store, a grandmother and her grand-daughter came into the store. After the elder woman played a few lottery numbers, the young girl (about 5-7 years old) approached the counter (with candy below at her eye level) and said to me, "What do you do when you run out of candy?"

Me: "We Get More."
Young Girl: "Yeah, but from where?"
Me: "Well, we order it from the Candy Manufacturer... like Willy Wonka."
Young Girl: (defiantly) "Yeah, but he's not real!"
Me: (taken aback) "Well young lady, it was just a metaphor... an analogy... an example... of someone like Santa Claus or the Easter Bunny."
Young Girl: (triumphantly) "Yeah, but they're real!!!"
Me: "I stand corrected."

As she walks out of the store (with candy in-hand), her Grandma says, "Just like the Tooth Fairy."

Ahhh... outta the mouths of babes. ;-)

Student Goal(s) for this Week:

Start to dispel the illusions you may have about engineering and whether it's really the right career path for you. It you decide to accept this mission, of becoming an engineer, this book will provide "just the gist" of what to expect.

Professor Goal(s) for this Week:

Stress the importance of learning technical communications because "poorly written equals loss of sales" for you... and for your company.

Also since it's the beginning of the term, consider taking 10 minutes for everyone in class to shake hands and introduce themselves. Then after, initiate solicitation for what the students thought of each other's handshakes without naming names or pointing fingers. A good handshake is crucial as these young students march forward. Reinforce that a good handshake isn't a contest... but that it should be firm.

The Benefits of this Stage in the Engineer's Journey:

You really don't know what you want yet... so you still have the freedom to roam.

The Downside of this Stage in the Engineer's Journey:

You really don't know what you want yet. Also, this is really the time in your college career when you're getting all your basics out of the way... such as calculus, chemistry, physics, and various humanities electives. So you can go a long way down the engineering path and not really know yet if it's really for you.

Even if you struggle, just making it through is important. That degree matters. While some consider it "just a piece of paper," it's real importance is that it's a piece of paper conferred that shows you are moldable, teachable, and able to jump through the hoops sent your way. That's why a degree really matters. It separates wheat from chaff.

If You Don't Know Your Destination, How Will You Know When You've Arrived?

Just realize that going to an expensive school may be impressive, but it can also burden you with a lot of student loans. That's where it matters to weigh whether living on campus is a better trade than choosing a commuter school.

It's up to you to get the best education you can muster.

Great schools provide firm and fertile soil for success, but it doesn't mean that the seed inside your soul is meant to be planted at the first school you choose. Sometimes, you have to make the most of where you're at… get great grades… so you can leap frog into graduate school at an elite university. That's the path I took. I chose going to a commuter school that offered me a 4-year full-ride scholarship, earned a 3.96 GPA, and then got into the university I really wanted to attend.

A commuter college doesn't offer the same "college experience." However, living at home while attending school saves you a lot in terms of student debt later on in life. It also alleviates issues you might have with bad roommates. Housing affordability issues won't be the headache they often present. On the other hand though, campus convenience is a beautiful thing, especially when you need its resources… such as a book in the library that might help solve an elusive homework problem.

Top 5 Things to Do at the Start of this Phase of Your Life:

The first coolest thing you can do in your first two years at college is to immerse yourself within "college life" such as going to an early fall football game when the weather is still nice and before classes start getting tougher later-on in the semester. What this does is introduce you to the school's "Tradition"… which is a crucial component in how you view, savor, and perceive your "college experience" when you look back in hindsight.

The second thing you should do is to consider joining a club or two at your school… such as an engineering society… which can lead to later networking opportunities. This shouldn't be confused with Greek Life in a fraternity or sorority though. The whole point is that joining an engineering society and being an active officer in it looks great on a resume when it comes closer to graduation. It can also lead to a summer job or two.

The third thing I'd do is to try limiting your partying experience during these first two years. When you go to college, a certain sense of freedom away from your parents is achieved, and alcohol often is easier to come by. The problem with this is that it can cause your grades to suffer. An example of this comes from my best friend from high school. He was awarded a four-year full-ride scholarship to a fine engineering college and lost it after his first year because he wasn't able to maintain a 3.0 GPA… due to starting to have his first "intimate" relations with his girlfriend, partying, and working a minimum wage job at an electronics supply store. Those extra-curricular "activities" detoured the focus that got him good enough grades in high school to earn a full-ride scholarship, and it cost him thousands of dollars in tuition money he'd later have to pay… as well as a much lower paying job offer from an engineering firm after he graduated from college. So the freshman-year partying definitely did have a trickle-down-economics effect on his life later on for years and years to come. Party wisely.

Fourth, find your way to your school's gymnasium. It's a great place to start making new friends at school… which definitely has an effect on your overall enjoyment of college. That's one of the scariest things any student faces when they come to a new school… not really knowing anyone. It's a frightening aspect of getting a new job in a new state… which means moving away from the roots of "home." Also, getting some exercise in the gym or shooting a few hoops on the basketball court is a great way to get some healthy endorphins flowing… which will help promote a better outlook on life in general too.

Fifth, start looking for an on-campus part-time job. It helps you find your college's jobs board, and school jobs are a lot more conducive to being able to do some homework in its boring-time nooks and crannies. Also, the people who hire you in these campus-jobs totally understand that school is your first priority… and not the other way around… like you might learn if you get a job off-campus… where their business is the top priority.

Top 5 Things to Consider at the End of this Phase in the Engineer's Journey:

First:
A commuter student versus living on campus… brings about the quest for personal space. As you age, you'll need to learn boundaries in relationships. This might be hardest for your parents. When the child starts to leave the nest and turn into an adult, separation anxiety issues can arise. The whole point of understanding how to respect boundaries is that they enable us to be considerate so the relationships we develop tend to endure.

Second:
Knowing the basics, the fundamentals, and the rudiments by heart matter. This is especially the case with math during this stage of the game.

Third:
Another thing to consider is whether you want to go into actual engineering or whether you're satisfied with just earning a degree from the school of technology. A "technology degree" doesn't offer as many opportunities as an engineering degree. It's the difference between a technician and an engineer. An engineer earns more. So that's why it's worth putting yourself through the math-intensive grind that makes up your first two years as an engineering student. In my opinion, I consider the technology degree to be one which is "math-light." It's not as rigorous… so

if math turns out to be a weak-point, you might want to consider a degree in technology instead.

Fourth:
These days, computer programming skills are growing more essential. While it may not be your cup of tea, just know that the more software and programming skills you learn; the better of a position you place yourself for the future.

Fifth:
Do your best. Even if you don't get the best grades, some CEOs of start-up companies didn't have the greatest grades or test scores. What they did have was passion... and vision. That's key... seeing a niche in the marketplace and taking the chance to fill it.

Lecture Ideas for the Students in Your Classroom:

Student attrition is a fact of life. Some aren't cut-out for the grind of college life.

Share this example from my freshman roommate's life. He went on to be an architect.

When he enrolled, they gathered all the architecture students into one large auditorium. It was packed. The chairman of the architecture program... which was notorious for being tough to complete all four years... shared the following. He said, "Look around. We have a jam-packed auditorium. Next year, only 70% of you will remain for your sophomore year. In your junior year, only 50% of you will still be here. For your last year, only 20-30% of you will remain and graduate... even fewer with honors."

Some Homework Suggestions for this Week:

Encourage the entrepreneurial mindset right off the start. Ask your students what they'd like to change about the world around them – no matter how large or how minute – and what they would do or build to change it for the better.

As one university describes, "Entrepreneurship is a mindset characterized by risk-taking, experimentation, learning from failure, challenging oneself, and seeing problems from multiple perspectives." Thus, it is about the ability to develop and implement new ideas.

My Ruminations about this Stage in my Life:

The most special part of these first two years in engineering school was that I met my "first true love" in life. While I'm glad that she wasn't the last (or the only) love of my life, it was special. While romance was definitely a part of it, what made it even more special is that we both shared similar scholastic goals. She was a straight-A physics student, and I was a straight-A electrical engineering student. That's important in any relationship... and it sets a firm foundation for love to take root. We met at a Society of Physics Students meeting in my sophomore year... on campus... so we understood the burdens that school places on a student in order to get good grades. That's important.

Another thing I found important during this stage was to network with enough older students so they could provide accurate feedback about what professors to take and which ones to avoid. It's a fact that certain professors are terrible. I had my share, but it was more so during my first two years of college. Some are easy. Other professors are tough.

This was especially true of the person this textbook is dedicated to – Zaven Margosian.

He was a genius at numerical calculus... which dealt with solving math problems by writing computer code. So when one of our department's math teachers had a serious illness, Zaven took over our class. At the time, he was the Dean of Arts & Sciences, but since his expertise was in our class, he juggled professors to cover the unexpected vacancy. At the time, our class was packed... because the teacher initially teaching it was notoriously great at explaining math and making it seem easy. Once Zaven took over, students began dropping the class left and right.

During that class he had a motto he'd repeat almost daily, "Think, analyze, compute, verify." One day, he asked our (now) half-full class what the motto of our class was. From the back of the room, one student shouted out, "Drop!"

Everyone laughed; however, those who remained for the entire term felt like we earned our numerical calculus education, and it became a rite of passage for all of us.

By earning good grades in math, an opportunity to tutor was offered, and it was a great move to take it. Tutoring definitely improves your own skill as you explain it to others. Also, getting paid to do something you're good at is a nice feeling early in your career.

This Week's "Take-With" for the Professor:

This time of the term is really a "getting back in the groove" kind of thing. The first week of class is like "getting your feet wet" first before you really dive in… so try to make this class fun. My feeling is that it's a class that should be an easy-A. That way, your freshman students get their grade-point-average off on a good note. What would warrant a grade of a B or a C? Apathy, absences, and tardiness would. The whole point though is that this course and this book are all about providing perspective.

That's because being an engineer isn't always someone's cup of tea.

Yes, it's a respectable career. Yet it may be more important for their parents than it actually is for the student. Becoming an engineer is a safe and secure route, but rarely does it lead to outstanding excitement that sizzles others with their stories. That doesn't mean it's absent though so make sure to inspire your students, but make sure they realize the actual realities. Ask them to entertain all the angles.

This Week's "Take-With" for the Student:

Orson Welles told Ruth Warrick when he was making his epic film *"Citizen Kane,"* "Find your heart's desire and follow it."

Moral of this Week's "Story"

Brevity is a great charm of eloquence. Learn this art: being brief.

All too often, engineers are notorious for providing way too much information.

When attending a STeM meeting (Science, Technology, and Mathematics) that was meant to introduce junior high school and high school students to careers in science and engineering, those engineers who volunteered gave overly dense presentations that provided too much information. I expressed my feelings that these presentations were communicating too much detail for the target audience age.

Immediately, an elementary school principal sitting next to me nodded and said,

"Less is always more."

Or said more elegantly, with one less word, *"Less is more."*

Week 2:
An Undergraduate Engineering Student's Last Two Years

Some Brief Brainstorming:

Seeing the invisible and making it a reality… this is engineering.

What is essential is often invisible to the eye… like "dark matter" in physics.

Some interactions in the universe on a cosmic scale don't add up with the mass of visible matter alone. It's why scientists hypothesize and propose the presence of "dark matter."

Find solutions and cures for a problem… there is a difference.

The Visionary… the Seer... doesn't solve repeatedly… they cure once… permanently.

The Humor in this Stage:

As you go further in your area of expertise, it will almost become second nature to speak in acronyms, but the funniest one I learned came from a female engineer who was fed up with her boss asking her how to use a piece of software. She told me, "RTFM is my favorite acronym." When I asked what it meant, she said, "Read the f***ing manual."

It's amazing how many people avoid this step… reading the manual.

The world would be a better place if more people did so.

Student Goal(s) for this Week:

Trust, but verify.

Think. Analyze. Compute. Verify.

Professor Goal(s) for this Week:

Teach them a firm understanding of "The Scientific Method" because it will serve them all their life and can apply in many different arenas in life.

A good engineer learns to make metaphors that enable a firm understanding by one and all. Reinforce the importance of being able to communicate their expertise to someone who has no understanding in the area. In other words, be able to explain it to someone as if they were a 5-year-old.

Being that engineers tend to be data-driven, they have the reputation for serving up information overload with detailed minutiae that makes it difficult for their listener to see the big picture… this is especially true if they don't possess the same context for understanding as the engineer explaining it to them.

Thus, also reinforce the importance of eloquence and brevity… and forming gestalts.

Being able to distill the quintessential essence is an art form. The sooner your students learn it as they proceed past their college learning; the better off they'll be.

The Benefits of this Stage in the Engineer's Journey:

A lot of "hands-on" lab courses where you get to work with the actual equipment.

You'll start to get opportunities for tutoring and internships… take them.

The Downside of this Stage in the Engineer's Journey:

As a lot of you younger students turn 21 years old during this phase, remember that life is full of sinusoids… with peaks and troughs… where we're advised to maximize the peaks and minimize the troughs. Yet the main thing to always keep in mind is that "minimizing the troughs" doesn't mean minimizing the pain via substances… as they can later lead to habitual addiction that makes those troughs a lot deeper later on.

It's one the things your parents are probably most concerned about… trying to help you avoid life's pitfalls. Like leading a horse to water though, as much as they may overburden you with their counsel and life-earned wisdoms, "They can't ride the bike for you." Life's "bike" takes a lot of balance that is constantly readjusted… all through life.

The problem becomes… at this stage in your life… parents can stifle and suffocate you with what is called "helicopter parenting."

Remind them that this is a time in your life when you need your own space to grow… into whatever tree your soul's "acorn" is destined to grow into.

If You Don't Know Your Destination, How Will You Know When You've Arrived?

I found it interesting that the very last class I took in undergrad was the one that made the most difference in my life… and it was an elective. It was "The History of Cinema."

Top 5 Things to Do at the Start of this Phase of Your Life:

First:
Don't contribute to classroom distractions (like cell phones and watching videos).

Second:
Remember the saying, "Small moves."
You can say it another way, "Babysteps."

The whole point is that you need to master one thing, then move to another… and then master that. Make a small move – well. Then make another – equally as well.

As Occam's Razor asserts, "All things being equal, the simplest answer has to be the right one." Or, it tends to be the right answer.

As my drum teacher insists, "Focus on (thee) One thing." He then added, "A jack of all trades rarely tends to be a master of one."

Third:
I'll say this a lot of times in this book. A good handshake is important. It leads to an unspoken communication between whomever you shake hands with. Make it memorable. The way you do this? "Firm handshake… but it's not a contest."

For example, this week I was having my car serviced and was offered to have it purchased by the dealer. In the end, I weighed all the angles, but the one thing that led me to decide to keep my car was the handshake I received from the salesman… as well as the general sales manager. Usually, salesmen are notoriously known for having a great handshake that can seal a deal. In this case, it was the opposite; the lack of firmness in their grip as they shook my hand had me feel as if something was amiss with their offer.

It's just one of those intangibles that can sway you from going one way or another when you reach a crossroads and are faced with a decision.

Interesting tidbit is that I was offered a lease of a new car that was loaded with options. That night, I solicited a lot of opinions about leasing versus buying, and the answer I liked best came from a young female entrepreneur who said, "The Millionaire's Mentality of buying a car (over leasing) is that at the end of the term (of your loan), you own the car."

Just some food for future thought…

Fourth:
Being that this textbook is about "the journey of the engineer," I overheard a quote by NBA hall-of-famer, Kareem Abdul Jabbar," who said (about "the journey"), "A lot of times… you don't understand it as you're going through it."

I tend to agree. Usually, it's only with hindsight that things make sense. Choices you made and their outcomes will finally make sense. Karmic boomerangs will arrive (for the good and for the bad) that mirror actions and choices from your past. A lot of times you'll have doors you want to open that remain closed. For those cases, hindsight will usually show that it's a case of pushing too hard when the door really said "pull."

Yet these moments of illumination create perspective. Perspective enables better decisions to be made as the future unfolds. The whole point is to make the best decisions you can using truth, grace, ethics, and integrity. The sooner you learn to do that now; the easier life will be later on. We all fall into the trap of lying from time to time. The more you lie though; the more you have to remember your lies. You might be surprised how many times people have wanted to hurt me physically… only to say, "But you're telling the truth… so I'm going to let you slide… as much as I might not like what you're saying." As the old adage goes, "The truth shall set you free." It does. Use it – always.

When you're "going through it," it's easy to get absorbed and immersed in the minutiae. That's where hindsight enables you to see life's mysteries that you face with clarity. You see that there's a science to it… a logic that applies… which makes it all make sense. While you're going through it though, sometimes you're too close to the details at hand to see that you're really in a hole and stuck. It's harder to discern the mountains from the molehills while you're passing them by… especially if you're still too young to have enough experience to ever have seen and witnessed a mountain's majesty before.

26

That's where wisdom comes from experience.

That's also why I write this textbook and hope it will become a college course at every engineering school in the world. It tries to present the context and perspectives you'll need before making the journey of the engineer to see if this mission is really the right one for you. While you're young, you may not know what you want to become yet. You just may decide to educate yourself in an area of study which you're good at. That doesn't always equate with what you love though.

As Pablo Picasso is said to have said, "If you do what you love to do everyday, you never work a day in your life."

Fifth:
Try to take a few engineering courses in other disciplines than what you're majoring in... sooner rather than later. That's because you may find a particular aspect of engineering to be an area that you take to like a duck to water. This happened to me with thermodynamics. I took it in my last semester of undergrad. I liked it so much that I often said, "If I'd taken thermo sooner in my college career, I might have switched majors... from electrical to mechanical engineering."

Interestingly enough, 30 years later, I was offered to teach a course on thermal and fluid sciences at the university level. So my enjoyment of its content, years prior, enabled me to enjoy presenting its content, years later.

Therefore, you never know how components from the past are actually fulcrum ingredients for the future. As musician Frank Zappa is quoted as saying, "A mind is like a parachute. It doesn't work if it isn't open."

Top 5 Things to Consider at the End of this Phase in the Engineer's Journey:

First:

One of the greatest tools you need to add to your engineer's toolbox is to understand what "The Scientific Method" really means… and how to put it into practice.

In essence, it really is a deductive process. You rule out a lot of extraneous ideas that aren't the real solution so you get closer to arriving at a solution that really works.

In other words, you have a theory, develop a hypothesis, test it, gather data, and then verify that it wasn't a random result. Consistent repeatability is crucial in engineering.

Second:
Tomorrow is not promised. Get as much done as you can today. Then, if and when tomorrow arrives, repeat. Get as much done as you can again. When you work that way, you never go to bed restless, and you eventually transition into a perfectionist who can meet deadlines… which is a trait boss's love… as it is the essence of reliability.

When people know they can rely on you, you become a very precious commodity.

Being trustworthy is a trait that will serve you well all the years of your life. Add it to your arsenal now. If you do, your future will be so bright; you'll have to wear shades.

Third:
Understand that by getting your engineering degree, your parents will probably expect "a certain kind of success" from you. However, that may not equate with what makes you happy once you get your first "real job" as an engineer.

For a good friend of mine, we met in a "History of Cinema" class during our last term of school before receiving our bachelor's degrees. Being that we were both film buffs, that summer we went to the movies a lot. We turned from being good friends to great friends along the way. That's when he confided in me that he wasn't sure engineering was for him, but being that he still lived with his parents who paid for his education, he was afraid to disappoint them by wanting to take his career in a different direction – teaching.

The result of that conversation was that I went with him to his home and lobbied on his behalf to enter into another university's "school of education" in order to get his teaching certificate… which is a requirement for secondary school teachers.

His parents were tough… but fair folks… and they replied, "We love you, and we support you… but we just want you to make up your mind. If this is what you really want to do, we'll back you up. Just make sure it's what you really want."

He nodded, and I replied, "Honestly, I think your son is going to be one of those memorable teachers whose students will be glad to have had and will remember all their lives with an air of gratitude."

As it turns out, he met his future wife at that next university's "school of education" as both were getting their teaching degrees together.

Over 20 years later, I visited him one chilly January… unannounced. Fortunately, his wife and four children were out… so we sat and shared some chit-chat. That's when he paused to show me two letters he just received from two of his students… right before Christmas break. Apparently, he hadn't shared them with his wife… for fear she'd be jealous (as a fellow teacher). Those two letters brought tears to my eyes. They expressed how happy they were to have my friend as their junior high math teacher and how easy he made math seem for them. They expressed how great of a teacher they thought he was. It was a special moment for the two of us… especially since he felt I was the only one with whom he could share them who had the context of how they originated… so long ago at his parent's kitchen table… when I stood in front of his parents and took the brunt of their exasperation over "throwing away his engineering degree" to be a teacher.

As it turns out, the engineering degree only made him a better junior high math teacher... simply because he had a wider palette of examples he could share as to where the math really applied and made a difference in the world.

You've probably been there. You've probably had those moments in school of wanting to ask, "Are we ever gonna use any of this stuff when we grow up?"

I guess the answer is, "You never know what the tide will bring."

It's better to know it and not need it... than it is to need it and not know it.

That's why we educate ourselves... to learn how to learn... and to teach ourselves whatever we set our minds upon learning.

Fourth:
Whatever happens next, be the hero of your own story.

The ingredients you may need are ambition, intensity, and the understanding that things will go wrong along the way from time to time. It's how you respond to what went wrong that matters and makes all the difference. Just remember that satisfaction may require multiple attempts. You rarely get it "just right" the first time around.

A metaphor which may help here is that in movies... at least in the good ones... every character "arcs" by making a change for the better... except the bad guys. They don't change. They don't arc. Thus, to be a hero, you must change and transform.

Fifth:
Begin to realize that people who possess talent don't feel a need to proclaim it. Said another way, "Beautiful things don't ask for attention." The best musicians I know don't need to tell me how good they are. They already know.

When I tell fellow musicians (especially drummers) of the time when I first met and was introduced to a very famous drummer in Los Angeles, I relate how I asked, "So I hear you're a drummer. Are you any good?" He just shrugged, scratched his head, and said, "Yeah, I guess I'm okay."

When I tell those musicians, who know who he is, that story, they all bust a gut laughing.

They usually then say something like, "When a player tells you that he's '*just okay*,' that's when you _know_ he's good!"

Lecture Ideas for the Students in Your Classroom:

Consider the concept of "parasitic losses." In general, it applies to electrical devices which are plugged in, turned off… but are still drawing power… AND… wasting money. Transferring that concept to engineering designs, the more streamlined you can make your design, the better it tends to be… especially as its iterations evolve into the future.

Discuss "Occam's Razor." It means that all things being equal, the most elegant design tends to be the most correct one.

Talk about "dimensionality" in relation to the book by "A. Square" called "Flatland." It tackles a two-dimensional world… until a three-dimensional sphere passes through its plane of existence… first as a dot, then as a circle of increasing circumference, reaching its maximum size at its equator, and then diminishing to a circle of smaller sizes until it becomes a dot again and then disappears. That "Sphere" then pulls "A. Square" out of its plane of existence to show that it is actually "A. Cube."

Some Homework Suggestions for this Week:

Based on the dimensionality concept discussed in Flatland, have your students use six (6) equal-sized lines to create four (4) equilateral triangles. Have them draw a diagram that shows the solution. Hint: the solution requires a 3-D approach.

(Solution: it's a three sided pyramid… a lesson in dimensionality.)

Another homework idea is for your students to research the Hadron Supercollider in Europe and discuss how it could lead to us discovering other dimensions. Then have them relate it to "string theory" in physics… which actually requires additional dimensions in order to make sense.

Additionally, ask them transition that short essay discussion into proposing what will be available 100 years from now.

My Ruminations about this Stage in my Life:

As someone who got two electrical engineering degrees, I often joke with people that if I had taken thermodynamics earlier in my college career, I might have decided to become a mechanical engineer instead… simply because I took to it like a duck to water. For me, I took thermodynamics in the last term of my senior year. By then, it was already a done deal, and my fate was sealed that I was firmly on the path to become an electrical engineer. The thing is; I was great at math… but that may not amount to what you love.

The whole point to living a happy and satisfied life is to find your heart's desire, educate yourself how to be an expert in that field, and then follow it. See it through. Even if it turns out to be the wrong path, at least you'll have seen it through enough to have the knowledge and perspective of coming to the realization that it wasn't really for you. That's valuable knowledge. Also, you'll never regret having at least tried.

Now with that said, let me get on to probably the most important thing I did during my last two years of my undergraduate engineering education. It was to become active as the vice-president of the student chapter of the Michigan Society of Professional Engineers.

While any engineering society activity is a good thing for networking and for your resume, our chapter was great because we offered educational "refresher" courses for the EIT (Engineer in Training) exam... which is the first exam you must take in order to become a Professional Engineer (a P.E.). Our university's Dean of Engineering was also a P.E. and made it a requirement for graduation that every engineering student had to take and pass this EIT 8-hour exam during our senior year. Since our student engineering chapter offered these refresher courses and charged money for them, we became one of the richest engineering societies on campus, and that enabled us the freedom to attend many National Society of Professional Engineers (NSPE) meetings in some really great destinations such as San Diego, California and Mobile, Alabama... both, right on the sea.

Further, being active in these Professional Engineer meetings enabled me to meet a lot of engineers who offered me summer internships and who provided me with great recommendations for graduate school and a plethora of other jobs.

Just like in any business, often, it's all who you know.

That's where being active in extracurricular activities can pay dividends in school.

Also during this stage of my life, I happened to go to the film theater with a bunch of foreign exchange students from Germany, Belgium, Sweden, and Finland. We were all doing summer research internships in 1989 when the movie titled "*Young Einstein*" was released. What a great film to see at that stage in my engineer's journey! I'd just turned 21 years old, and that movie starred a young Aussie named Yahoo Serious whose spoofed Einstein character was trying to split the beer atom. For bunch of math and science geeks, it was a great film to see which bonded us together into better friends.

If you have a chance, try and find this rare cult classic and watch it while young. It's the perfect film for this science and engineering stage and at your age too to boot.

This Week's "Take-With" for the Professor:

Education's purpose is that it is an activity that can get us to a better place. The point isn't trying to get good grades in order to impress Mom or Dad. The take-with is that education should become important to the student. The inner-fist-pump moment of achievement isn't one that's given. It's one that's earned.

This Week's "Take-With" for the Student:

As you depart from your college with an undergraduate engineering degree, you're going to be walking into a world where "planned obsolescence" is a very real reality. It means that products purposely get produced to eventually need an upgrade… so that consistent purchases are made which keep that company alive and able to thrive. After all, your best customer is a repeat one.

Durability often equates with quality, and the saying that applies is, "They don't make 'em like that anymore." It's true. Decades prior, things were designed to last longer. As we evolve into the future, things are trending towards lasting shorter. It's consumerism.

Try to keep costs down while making quality better than before.

Even if you, and your skills, become obsolete some day, doing things with quality will always benefit you in the long run.

Moral of this Week's "Story"

"Nothing happens until midnight, and nothing good happens after 2 a.m. Kind of leaves a narrow window of productivity, eh?" – Overheard in the Wilson Student Team Project Center at The University of Michigan.

Remember, the importance is that education is important to You.

Week 3:
Graduate School Decisions

Some Brief Brainstorming:

Studies show that stress tends to jump in autumn as school starts.
More headaches are reported at this time of the year. Some
suggestions on how to alleviate it are to limit screen-time… and
also to go slower. Learn to pace yourself. Life is rarely a sprint.
It is more often a marathon. Grad school is one such marathon,
and it tends to push you to your academic limits unlike any other
experience in your life.

The Humor in this Stage:

On the cover of an old math journal it had the cute quip, "So
what's your problem?!?"

Additionally, in the movie "*Contact*," a series of scenes take
place at New Mexico's VLA (very large array) radio telescope
observatory where a bumper sticker is seen in the background of
one of the offices which says, "Astronomy is… looking up."

Student Goal(s) for this Week:

Analyze the benefits of the typical two years it takes to earn a
master's degree and weigh it against the bang-for-the-buck you
receive with only a bachelor's degree. Then ask yourself if you
feel you'd be able to get back into the "going to school" mode of
living, working, and studying if you were to go to work for a few
years and then want to come back for a master's degree later.

Professor Goal(s) for this Week:

Encourage the concept of technology transfer at all stages of
college studies so that all of the ideas born inside the university
environment find their way off the page and technical papers so
they find helpful use in alleviating the needs of society.

The Benefits of this Stage in the Engineer's Journey:

If you do decide to get an engineering job immediately after getting your bachelor's degree, a lot of times, your company will pay for you to get your master's degree.

Graduate school matures you further and develops your degree of expertise in the field of study where you wish to specialize.

Also, it enables you to enjoy the academic environment in a deeper degree. It calibrates and refines your expertise into a jewel-like prism with kaleidoscopic proportions and possibilities which await ahead in the research whose waters you'll wade into.

The Downside of this Stage in the Engineer's Journey:

Once you reach this stage of your life, the level of difficulty definitely intensifies.

Just be prepared for the challenge. Grad school was the toughest two years of my life. During my first month of grad school, it was so grueling that I thought of quitting. I'm glad I didn't. Besides, no one likes a quitter.

If You Don't Know Your Destination, How Will You Know When You've Arrived?

If the brain is like a muscle, in grad school, your goal is to develop "muscle memory."

Muscle memory is the essence that you achieve after doing a task so many times that it becomes second-nature. It turns into a skill where you don't think; you just "do."

In sports, sometimes the mind gets in the way. In baseball, a batter can over-think himself into a strikeout in a heartbeat.

Developing "muscle memory" is the essence of learning how to get into "the zone."

"The zone" is that magical place where you're functioning on all cylinders. In bowling, getting into "the zone" eventually earns you a perfect 300.

In engineering, it's being peaceful when you should be stressed. It's being collected in the midst of chaos. It's being forgiving when people expect you to be angry.

Reach this state of mind, and you'll always be at your destination… no matter where you go… simply because you'll be the soul of every place you're at.

Top 5 Things to Do at the Start of this Phase of Your Life:

First:
Just know that your parents spend a lot of money for the education of their children, and after they do, they expect "a certain kind of success."

It may not be the kind of success that makes <u>you</u> happy though.

If you change course, the main balance to achieve is to know they need to be happy that you're happy. Therefore, reinforce to your parents that "success" is not a destination. It's a journey.

Second:
Be grey. Be black & white. Be colorful.

Be each color of the ROYGBIV spectrum for at least one moment in time. You never know. You may find one which fits and suits you best.

Third:
Fall in love with the academic environment.
It fosters a spirit of exploration.

Fourth:
BACK-UP YOUR WORK. You can't imagine how many times
I kicked myself for not backing up my work on a consistent basis.
Really, it has to become a discipline. Sometimes, I'd be working
on a homework assignment, and the power would go out. Other
times, something happened to my computer… making it useless.
At times, there were files I really wanted to retrieve, but my non-
functioning computer made it impossible. One mentor of mine
advises using an external hard-drive at all times… especially if
the computer is a desktop version (and not a laptop).

The earlier you make backing up your work like a daily (even
hourly) consistent "yoga," the better off you will be in the long
run. Make it a ritualistic routine for peace of mind.

Fifth:
In grad school, you should ask two simple questions, "Why?" and
"Why not?"

When it comes to perceiving and envisioning your future success,
ask "Why not me?"

*Top 5 Things to Consider at the End of this Phase in the
Engineer's Journey:*

First:
Grading on the curve in difficult courses is often a necessity in
graduate school simply because major research institutions don't
like to have failing students amongst their graduate ranks. It
hurts their national perception. So the lowest grade a lot of
universities will hand out will be a B-minus. They don't want
drop-outs. They don't want failures. If you've gotten accepted
and gotten this far to go to grad school, they'll work with you and
find a way to "get you through." The whole point is to make a
concerted effort… also known as "the ol' college try."

Second:
There's an unfathomable scale to the knowledge available in the
universe. Even the term "universe" is a poor descriptor when it

comes to cosmic proportions because its prefix of "uni" implies "unity." When you look at the stars and galaxies, there are many universes… of which ours is lovingly referred to in popular culture as "The Milky Way."

In graduate school, you will be expected to try and tackle a small chunk of that "unfathomable scale" of knowledge. You may even become its foremost expert.

One paradigm that changed for me with grad school was the perception of the order of dimensionality. That's because in many advanced schools of study, they consider "time" to be the first dimension… and then the x, y, and z dimensions of length, width, and height being the second, third, and fourth dimensions.

Additionally, some hypothesize that "consciousness" is the fifth dimension and that "historical consciousness" is the sixth dimension… since if we don't remember our history, we will be doomed to repeat it. After all, insanity is doing the same thing over and over again… and expecting different results each time around. That only leads to circular paths… instead of spiraling and intertwining paths like the strands of DNA which, together, form a complementarity.

Third:
Graduate school enables additional career opportunities as well as greater salary offers; however, it also shows companies looking to hire that you are trainable to a greater degree of refinement. That's something an engineer told me when I first started working in a large corporation. She said, "We could train a monkey to do this job, but we require a master's degree because we want an additional level of expertise and understanding of the underlying science behind and beyond what we do because we never know when we hire someone who our next Einstein will turn out to be."

Fourth:
A goal without a plan is just a wish. If you want to achieve your greatness, what is your plan to achieve it? My example is to list

three musical prodigies and then make the metaphor how each one is like another step up... similar to going from a bachelor's degree to a master's degree to a doctoral degree. Those three music composers (for me) are Mozart, Beethoven, and Wagner. Mozart was the first prodigal genius in that list, but Beethoven offered a deeper amperage... and then came Wagner... who pegged the needle into the red (so to speak) with his *"Ride of the Valkyries"* which was made famous in the movie *"Apocalypse Now"* by Francis Ford Coppola.

Just as there was a progression of intensity and amplitude when we went from Mozart to Beethoven to Wagner; the same progression applies to the degree of refinement you'll achieve as you gain a bachelor's degree... a master's... and a doctorate.

In a sense, it's like combing out the knots in one's long and tangled hair. Each pass of the comb refines the hair to a sense of smoothness. However, each step in that educational progression's series is a deeper commitment. The conundrum comes if, years later, you decide that engineering isn't for you. Those degrees offer a "perception of success" from fellow peers, friends, and families. If you "throw them away" and pursue another passion later on in time, many will question your sanity. Many will wonder why. Many may even estrange themselves from you if you someday make such a decision.

The whole point is that what you're good at may not equate with what you love to do.

It's a valid consideration because graduate school offers a minimum of two years of additional education in order to achieve another "refined" degree.

The choice is yours... should you decide to accept the graduate school mission.

Fifth:
In Hollywood, after a film is finally finished and released to theaters, the saying that applies is "Next!"

It implies that you can only rest on your laurels for so long. Often, you're only as good as the next project you're working on. People not working towards a positive direction are often perceived as good-for-nothing derelicts... or bums.

"Stillness" is a good thing. Meditation helps calm frazzled nerves. However, if you have ten oars in the water and are trying to row each one, you just may row yourself in a circle.

Lecture Ideas for the Students in Your Classroom:

Stage a "mock debate" in your classroom by soliciting your students (via a show of hands) with how many think that they'll eventually go on to grad school. Have them debate why they should go to graduate school and why they shouldn't go.

Another idea is to have them consider what happens if they get accepted to the graduate school of their choice but aren't offered any financial assistance (such as a teaching assistant or a research assistantship).

Also discuss your own personal motivations for achieving the degrees you received and what you felt (as a professor) that they enabled in your life.

Some Homework Suggestions for this Week:

In writing a screenplay and directing it to film, there's a science to closing a scene... to buttoning it... with a hook... that pulls you through and invests your interests into wanting to learn more. Therefore, they present "the hook" by ending on "the look."

In other words, you don't end a scene with words. You end it with emotion. You end it with expression. A picture says a thousand words. It's like trying to create the perfect "famous final scene" that lasts with audiences for years to come.

Likewise, when you're deciding on graduate schools, the best thing you can do is to visit the campus before making your decision on which one to attend. Is "the look" of the campus one that would "hook" you into attending it? Is it an academic environment you can easily envision yourself working within?

Therefore, have your students visit two nearby college campuses and their websites. Then, have them form a list of "pros" and "cons" for why they'd choose one school over the other if they had to make a choice for which campus would foster the best environment for their future education to thrive.

The whole point is that, in life, if you return to your old college stomping grounds, you want to look back at those days with no regrets. For example, attending a college with great sports teams matters because someday you might be in the stands as an alumnus rooting on your old school. It creates a feeling of belonging.

Thus, as an alternate homework assignment, have your students describe why they feel like they belong at the specific college they're attending right now. Have them describe things they like about this school. Have them explain a step or two they feel their school could do to be cooler than cool.

My Ruminations about this Stage in my Life:

When I look back on my decision to go to grad school, I now have 25 years of hindsight since that decision, and I can see how it was an ingredient to becoming the person I hoped to be… who also is the person I am now. It was an essential ingredient of the stew I needed to brew into the soup of my life.

As a grad school student, I learned how to learn. Thus, when I took over a decade's sabbatical away from engineering in order to become a writer, I took upon the task of teaching myself how to write… better than I had as an engineer. You learn that there are certain formulas that work. For example, movie screenplays have a proven formula. Likewise, writing a mystery starts with knowing your ending, and then working backwards when you scatter the story's clues... that way, when you arrive at the start of the story, you know that all you really need is one really good one (one really good clue) for your hero to get started. Then all you need is to be brave... at the right time and at the right place... for perhaps 10 seconds longer than the next guy... and then to let the chips fall where they may. That's because once you get started on that journey, it often feels like you've gone too far to turn back now.

Sometimes, in the end then, the hero realizes that the treasure (or the answer) they searched for all around the world... was really in their own backyard. What they gain by making the journey... are the wisdoms they gain and the sights they see along the way.

This Week's "Take-With" for the Professor:

Communicate with a snowshoe awareness as the term progresses... an awareness which distributes the weight. Remember, you're providing an overview of the overall journey.

Also, discuss the importance of studying for the GRE exam in order to get into grad school. Explain the three parts of the Graduate Records Exam. Since the GRE is required for grad school admission, offer ways they can maximize their scores.

The best way is to do practice problems out of a GRE Exam preparedness book... which explains the procedure (or *trick*) with which certain problems on the test must be solved.

This Week's "Take-With" for the Student:

You can walk on water. Believe it. It is possible. How? When it is ice, you can.

Graduate school calibrates seeing in this enhanced perception. Find each other. Embrace the diversity. Drink it up.
Have the courage to go into the unknown.

Moral of this Week's "Story"

Focus on the "One" thing.

Ask yourself what matters most. Focus. Manifest it into reality.

If that vision for your future includes grad school, focus on it singularly so it brings a bounty of opportunities later on. Also, consider going to grad school immediately after getting your bachelor's degree. It makes things easier because you haven't forgotten any of your school skills (such as advanced mathematics) yet.

<u>Week 4</u>:
Grad School as a Teaching Assistant

Some Brief Brainstorming:

Teaching labs

Teaching low-level courses

Each needs a different approach.

Getting the keys to open the classroom door

Make sure all the parts are available to <u>all</u> the students.

The Humor in this Stage:

In one lab course I taught, I got behind because it was my first term, and I didn't know that I was expected to get all the lab components for the students. Also, I wasn't given a key to the lab at first… so one student complained to the department's chairman on my lack of preparation. When I alleviated these issues and got the course started, by the end of the term, apparently I got ahead of the lectures being given by the professor… so that same "anonymous" student complained to the chairman again. Yet, if I hadn't started late, I would've been even more out ahead of the professor's lectures… so I was darned if I did and darned if I didn't. My take-with from the experience is that if you try to make everyone happy, eventually someone will find fault with our approach. Sometimes, all we can do is to try our best and then just roll with the changes.

Student Goal(s) for this Week:

Take the time to analyze the traits your favorite teachers have employed which made them your favorite teachers in the first place. Try to add them to your arsenal.

Professor Goal(s) for this Week:

If you sense disappointment amongst a student (or students), consider reversing gears. Consider an alternate approach that leads to their overall satisfaction. This may mean soliciting advice from your students as to how to teach better in a way which enables them to walk away from the experience with a smile. It also will calibrate your approach in the future so you can deliver on their expectations of the course and of you. It also may require pulling a student aside to speak about it in private. Sometimes, a student won't want to admit frustration in front of the others in the class. This personal approach is really just a solid step in showing your students that you care.

Then transition this approach to teaching them how to teach best.

Reveal some of the items in your teacher's bag of tricks.
Communicate what seems to work well in class.
That way, they know if they have to teach.

The Benefits of this Stage in the Engineer's Journey:

When you teach, you learn the material to a deeper degree because your job is to communicate its intricacies. In order to share its nuances, you must understand it yourself. As a teaching assistant, one of the benefits is that the classes you teach have a structure which is clear-cut. You know what time you have to teach. Once you've graded their assignments, you can get back to your own coursework at hand. This degree of knowing when you're "done" isn't as obvious as a research assistant.

The Downside of this Stage in the Engineer's Journey:

One of the difficulties of being new to teaching is overcoming your "stage fright" while in front of the class. It takes a while to grow your confidence in your approach. A lot of times, a school will give you a class and expect you to figure it out as you go. However, it's said that not admitting your nervousness can also come across as great excitement.

Admitting nervousness is akin to showing weakness.

If You Don't Know Your Destination, How Will You Know When You've Arrived?

The sharing of knowledge and know-how helps perpetuate the evolution of our maturity… as a person… and as a species in the human race. Pass along the knowledge you've learned first-hand because true understanding often comes in the "doing."

Top 5 Things to Do at the Start of this Phase of Your Life:

First:
Once you start teaching others, you realize how much you abhor tardiness. It's a bad habit. As I've learned, "If you're early, you're on time. If you're on time, you're late."

Taking that sentiment one step further, there's one film studio mogul who would say, "If you're not going to come in on Saturday, don't bother coming in on Sunday."

Second:
Know that there is a looming teacher shortage.

Ask yourself how much you like teaching during this stage. Being a teaching assistant provides ample practice. The thing that may surprise you is that if you give your students some latitude to have fun and to be creative in the presentation of their assignments, you may be thoroughly impressed. I was. While most students just turned in their assignments "by the numbers," a few turned in some really innovative and unique lab reports which made me proud I asked them not to be so "dry" in presenting their data and findings. This adjustment in their approach landed them more than a few A+'s.

Third:
Life has a lot of empty spaces. Being around each other is often the medicine which makes it all bearable. The fact is; people

need people. We need interactions amongst each other… in person… and not just online. Otherwise, the walls can close in.

Teaching is a nice thing. It opens avenues to students who have a fresh perspective. They see life in a different light. Those lights can be beacons in the darkness of seemingly endless homework assignments and studying. As numerous professors have said when considering retirement, "I'd miss the students. They keep me young. They bring life into my life."

As a teaching assistant, the age difference between you and your students may be minimal, but it offers an excellent opportunity to network as well as teach. That's because as you grow older, minimal differences in age begin to matter less and less.

Fourth:
When a person teaches, they learn to pass the torch of knowledge to a new generation of visionary thinkers. It's the foundation of becoming a mentor. This is where your educational know-how can guide students past the pitfalls of potential mistakes. It also enables you to educate and illuminate others on a one-to-one basis. While book-learning has its place in the development of wisdom, having a mentor to show another the ropes of how things are really done often necessitates a sitting or standing side-by-side so you can point out the methods which work with hands-on experiences. As the saying goes, "Show, don't tell."

So in essence, a good teacher is like a mountain Sherpa… someone who knows the safe path and can guide you to your destination. Be that "guide" for your students.

Fifth:
Many times when you're assigned to teach a lab or a course as a teaching assistant, you'll be assigned a higher-up professor who is in charge and calls the shots. A lot of times, they'll just expect to throw you in to hit the ground running without explanation.

If this is the case, try to find a past teaching assistant of the same class to show you how to perform the labs correctly. They may even have answers and solutions manuals which can make your job a whole lot easier. Make sure to inquire with your class' professor if such a past teaching assistant of the same class is available. They'll be a lot closer to the truth of how the course or lab is actually run with a healthy dose of nuts-and-bolts reality.

Any leader knows that getting data from afar is always helpful, but it doesn't match getting up close and personal. Be that kind of teaching assistant who has an open-door policy and is always willing to roll up their sleeves to help the student to the greatest degree as possible. Not only does it help the students, but it also sets a trend and a routine of excellence which makes you a consistent and dependable asset – anywhere.

Top 5 Things to Consider at the End of this Phase in the Engineer's Journey:

First:
Try to get feedback from your students on how you did and how you can do it better the next time around. If you solicit this response from your students, you may get different responses than if they grade your performance as a teacher anonymously.

What this does is help steer your ship if you're off-course and in need of an adjustment in the approach and methods you utilize.

Second:
Make sure your students know that you are available for providing them with written recommendations. In return, they may even provide you with a few recommendations and endorsements of your own.

For example, if you use a business social media networking site, such as LinkedIn, encourage your students to connect with you. For those who do, write each of them a recommendation. It won't take a lot of time out of your schedule, but by doing so unsolicited, you may be surprised how much your students appreciate it. That's because it enhances their online reputation and perception. It's the Golden Rule. Do onto others as you would like to be done onto yourself.

Third:
Sometimes at the end of a class, the department you're teaching in will ask you to return the materials they provided. If at all possible, make copies... especially of available solutions and techniques for correct completion.

An example of this is that I taught a lab which instructed students on how to properly calibrate a spectrum and a network analyzer for electromagnetic measurements. Six months later at my job within a corporation, having those notes handy enabled me to make valuable contributions at the company's far-field satellite antenna range. Since I knew how to use the equipment, thanks to my notes from being a teaching assistant, I became an immediately valuable asset during the time when a new-hire is usually still learning the ropes of how things are done at that particular place of employment.

While you never know how much of your schooling you'll actually use on the job, it's better to have those notes handy and not need them than to need them and not have them.

Fourth:
Did you offer extra incentives for your students to excel with you as their teaching assistant? Did you offer extra credit assignments? We all realize that getting good grades is an important goal. A good grade point average leads to better future job offers.

Also, did you teach by walking around and asking what specifics a student may be struggling with? Students are often shy about admitting their stumbling blocks. A good teaching assistant offers their assistance by making it easy for the student. Thus, you make it easier if you come to them instead of them having to resort to coming to you. Such a forthright approach often nips a lot of problems in the bud before the student goes off in an incorrect direction while attempting to solve a difficult problem or procedure.

Furthermore, the number of times you show your students how to do a problem correctly only reinforces your own expertise with each time you communicate the proper path. One teacher once admitted that although he'd taught a certain concept numerous times, it wasn't until about the tenth time that the light bulb went off for him so that he truly understood its complete nuances for the first time with an insight that had remained elusive in the past. He said that the students hadn't known about his eureka moment but that he recognized it when it arrived, and it fueled a more passionate approach.

Fifth:
Were you considerate? Were you as prepared and as consistent as you'd like to be? Were you reasonable with your students when they had minor issues arise? Did you bring good energy? Were you humble? Did you share the credit where credit was due? Did you take responsibility for any mishaps caused by you? Did you show belief in yourself, the course, and in your students? Did you truly listen when a student had a problem or a concern? Finally, did you follow through with your end of the bargain?

This list of questions highlights small things a good leader can apply to continue raising the bar in any pursuit. They apply to teaching assistants too. Ask them of yourself.

Lecture Ideas for the Students in Your Classroom:

Ask your students to think of one or two teachers who made an impact in their lives. Go around the room and have each one contribute what that impact was. Write them down on the board. Try to have each student volunteer a different trait they admired from their most memorable of teachers. Breadth and variety in successful approaches will illuminate one and all… including yourself. Teaching is an artform… not a "to do" list.

Discuss the importance for a teacher to stay a few steps ahead of the students while preparing for upcoming classes. Preparation is essential to good teaching.

Thus, this week's set of lectures is preparation for the potential possibilities that some of your students may find themselves in the role of a teacher at some point in life.

Some Homework Suggestions for this Week:

Ask your students to make a list of traits they admired in teachers from their past. Then have them also list the things they didn't like about how some teachers handle certain situations. The goal is the educate the student on the "do's" and "don'ts" of good teaching so that if they are faced with being an instructor at some point in time, they have it in their understanding what makes a good teacher great. Then ask them how they'd do things differently if they had to trade positions with that particular professor for a day.

My Ruminations about this Stage in my Life:

When I applied to graduate schools, I got accepted to two of the three I applied to. One of them just offered me admission into their school's electrical engineering department, and the other offered me a teaching assistantship. It was a letter of admission which doubled as a job offer. It listed the salary I'd receive, the health benefits they'd provide, and included free tuition. For me, it was a no-brainer which one to choose. Free is nice.

Once I accepted the school's offer for the teaching assistantship, an interesting thing happened. I started being solicited by professors within the department who were actively looking for research assistants. There were in-office interviews to see if there was a good fit. However, I decided to keep my teaching assistantship offer for that first term in graduate school. It was a fortuitous choice because during that first term, I started becoming a lot more familiar with the professors on campus and their research interests.

It enabled me to turn the tables on the research assistantship search. Instead of professors soliciting me (as in the summer before school started), I found the professors whose research interested me the most and solicited them if I could be their research assistant. After all, one of the requirements of a master's degree is writing a research-based thesis.

This landed me a research assistantship after my first term as a student and a teaching assistant was complete. However, after a year of research, I realized that I didn't want to become a doctoral candidate, and so during my last term of getting my master's degree, I worked with my department and reverted back to being a teaching assistant again.

In hindsight, that second class in my last term before graduating with my master's degree was a fateful stroke of good luck, because I taught students how to use equipment that I later used in my first engineering job within an aerospace corporation. If I didn't go back to teaching during graduate school, I wouldn't have had the valuable experience of using that lab equipment in my toolbox. So while we may not understand why we go through certain experiences life throws at us as we're living them, they usually make a lot more sense once some time has passed. My teaching assistantship was a case in point. It was a crucial component towards me making valuable engineering contributions in my first job fresh from school since I had just taught lab skills that I'd use daily months later.

This Week's "Take-With" for the Professor:

Consider why you became a professor. Brainstorm your likes and your dislikes about the job you possess as an engineering instructor. Communicate this list of pros and cons. For example, many professors I know abhor the department meetings and the bureaucracy, but they love teaching the students. Many others love having their summers off from teaching classes so they have the freedom to explore and enjoy their lives much more. Share these tidbits. Be honest. You never know which of your students may become "teacher of the year" sometime in the future, but if it's, in part, because of your contributions to their perception and excellence, you'll never regret it.

This Week's "Take-With" for the Student:

Teaching is a valuable and legitimate profession. Getting a teaching assistantship to graduate school helps pay your tuition and provides a salary with health benefits on top of it. If you're offered a teaching assistantship and a research assistantship to a university, consider taking the teaching assistantship first. It's more clear-cut than starting out doing research right-off-the-bat. Also, it provides valuable perspective as to whether a career in teaching should be on your horizon as a destination. You never know what you're going to like and enjoy unless you try it first. Being a teaching assistant allows such a valuable sample in the toolbox of your experiences.

Moral of this Week's "Story"

Ten mistakes you make today are ten things you do right tomorrow.

Sometimes, you just... want... to stay in the moment.

Great moments are precious.

Value each one.

<u>Week 5</u>:
Grad School as a Research Assistant

Some Brief Brainstorming:

In grad school, you'll get the feeling of being recruited by professors. I didn't consider writing computer code as one of my strong suits… so some professors passed me by after they interviewed me for the possibility to be their research assistant.

The Humor in this Stage:

The spirit of exploration is great, but it doesn't always produce tangible results you can place in a person's hand and show them you've been accomplishing work. This is perhaps especially the case with parental perceptions of your progress. Having something to show for your efforts isn't always available while doing research, and in some cases you'll feel like you're finding your way in the dark… or having to learn a new language… the language of the science you're on the forefront of.

Student Goal(s) for this Week:

Fall in love with the academic environment.
Fall in love with the exploration research enables.

Professor Goal(s) for this Week:

Examine the "publish or perish" reality at a research institution. It goes along with having something tangible to present to people… and fellow researchers. Explain that many professors get grants and must have something to show to justify such money.

The Benefits of this Stage in the Engineer's Journey:

Should the engineering student decide to travel this path as a research assistant, you may find it a bit idyllic. You are given permission to explore… to find the foremost frontiers of science and determine if it's a beach-head you want to establish as the core basis of what you will specialize in… should you continue on to your PhD. If you do, years of research will be required, and it will eventually become the foundation of your thesis.

The cool thing here is the freedom permitted in the art of exploration.

The Downside of this Stage in the Engineer's Journey:

While academic exploration's freedom is nice, its vast array of depth can make your efforts seem daunting… especially early on when you are still new to this foreign field of study. You might find yourself researching publications in technical journals where you have to decipher what fellow researchers in this field are working on… and initially, it'll probably feel like trying to decipher a new language. Since the writing and the science in technical papers is all about learning what others are doing at their particular frontier in this field of science, it might take hours and days to truly understand their findings. Then, you must translate and integrate that knowledge into your field of expertise so that you can later differentiate it out in the form of a thesis or a future publication with your research professor who is sponsoring your research. After that, you may get requested to present your paper with your findings to a group of peers from other research facilities, think-tanks, and university institutions. It is the forefront of "technology transfer."

If You Don't Know Your Destination, How Will You Know When You've Arrived?

It's hard for a research assistant to know what their actual destination is in terms of how much research to do. This is the reason some students stay for post-doctoral work after they finally receive their PhD… the field of exploration is still ripe with potential.

Usually, the destination here is getting your master's degree and perhaps receiving your doctorate if you do decide to stay.

A research institution hopes their graduate students stay for their doctorate. The school gets the most bang-for-the-buck if those students stay; therefore, there is a definite feeling of "discernment" you'll feel as you work towards your master's degree. It's almost an unspoken fact that you'll have to disappoint your professor who is sponsoring your research if you decide staying for the PhD isn't for you. It is a choice you'll be faced to make in graduate school... and it tends to be the hardest decision you'll make during those first two years.

It's kind of like deciding when (and if) to trade-in your car for a new one. Keeping the old car limits your debt ceiling. Getting a new car increases your debt responsibility. In grad school, getting a PhD increases your time responsibility you'll stay at that school.

Top 5 Things to Do at the Start of this Phase of Your Life:

First:
Realize that becoming a research assistant means that a professor will sponsor your tuition bill, provide health care benefits, and a salary stipend you'll live on. Thus, that professor will have expectations you'll need to do, perform, and fulfill in return. After all, you become beholden to the ones who hold your debt and/or provide for your needs.

Second:
Start the discussion of what your sponsoring professor will want to publish. The earlier you do this; the greater the time you allot for subconscious percolation of what to write.

For example, early in 2015, I was encouraged to write a textbook, but I had no clue what engineering topic I'd try to tackle. It took until the end of August 2015 to arrive at the conclusion that I should write about "the journey" rather than trying to attempt taking on a specific aspect or discipline within engineering.

The point is; you want to find a niche that needs to be filled by your contributions and expertise… then publish your findings. For this book, it took 30 years of "research" to gain the knowledge and perspectives provided here. I had to go to undergrad, grad school, and then work for a corporation, a start-up firm, a mid-sized company, and then for a university over that span of 30 years. While my stays may not have been as long as my friends and family thought I should stay, I endeavored in every strata long enough to fulfill expectations, get paid, and gain the perspective I now share here.

In hindsight, it was as if each level of engineering strata I experienced was a stepping-stone to cross the river of life in order to see engineering from another shore's point-of-view. Thus, each of those strata… which make up the chapters of this book… formed an overall gestalt three decades later. In other words, I now see a forest through the trees.

Third:
Realize that research is like crossing an ocean to arrive at a shore of understanding. You'll have to consent to lose sight of shore for long stretches of time. During this period, you may gain the feeling of being "adrift" in the sea of knowledge. This is where keeping your moral compass calibrated to knowing what you want to achieve via this stage's platform is crucial. Try to know what you want to get out of all this research. The answer to that quest may change over time. It's okay if it does. That's life.

Fourth:
Ascertain how often your research professor will want to meet with you. Also, ask if that professor has an open door policy where you can just drop in unannounced if you have a problem or a concern that needs to get remedied sooner than later. Some professors may just want to see you once a week for their powwow with you. Some will want it at a scheduled day and time. Others may be a lot more flexible.

For example, in grad school, my research advisor was very busy, so my meetings with him had to be shoe-horned into his schedule… where I had to have a list of concerns and questions I needed perspective on or answers for… all lined up. In the aerospace corporation I worked for, the senior scientist who mentored my work and my progress insisted his door was always open whenever I had a question or was stumped. His mentorship was more of a daily one without a structured schedule… and honestly, it was a lot closer to the grad school experience I desired at its start than what I received at its end when I decided to graduate with my master's degree and get an engineering job.

Fifth:
Typically, every engineering graduate school student is offered an office with a desk. Usually, it's an office shared by many similar grad students. Some will have cubicles. Others will have a more "open" floor plan which facilitates greater communication (and often, greater opportunity for distraction). Try to find an office that allows your work there to prosper. Lobby with the appropriate secretaries to place you in the office and at the desk which enables the environment most conducive for your research and homework to be completed as efficiently as possible. Sometimes, it will be the proximity to your classes you're attending. Other times, when you're done with most of your classes, you'll want to find an office with a convenient proximity to your research advisor. Often, you just may want to share an office with a friend who makes being there enjoyable. After all, having someone to talk to who shares your perspectives as a student is priceless. Research can be a lonely task, and people need people so the walls don't close in. Thus, having a good office on-campus can facilitate greater progress without the same distractions you may have at home, at your apartment, with your roommates, etc.

During one stretch of time during grad school, I had offices in two buildings. One was rarely used, so I didn't personalize it with pictures and personal belongings. Appropriately, I hardly did any work there. Likewise, I didn't view it as a place conducive to getting work done. It became more of a spot to put my stuff while I had meetings or business to attend to in that building. The other office was where I spent a much larger chunk of time during that period… so I personalized its environment with deeper diligence in order to make my time spent there as enjoyable as possible. Yet having both offered the opportunity for a change in scenery of my surroundings when I needed to make an escape from the place I was at. Having that option was nice. Sometimes, the distractions at the office I found "friendlier" and more "familiar" made getting any substantial work accomplished difficult. Sometimes, someone would want to talk when I wanted to study. Other times, the volume of people talking made it difficult to concentrate on my task at hand… so an escape to the quieter office became desired.

The whole point is that you're going to spend a lot of time doing research during this stage. Find and forge the environment which fosters your best chance for success.

Top 5 Things to Consider at the End of this Phase in the Engineer's Journey:

First:
Once you've done research at the university level, you achieve a valuable tool for your life's arsenal – you learn how to learn. You learn that you can teach yourself anything you set your mind on learning. Also, you earn an appreciation for research and academia.

Second:
Ask yourself if you want to continue your life's journey as a researcher. Is it your cup of tea? Do you want to take your skills to a think-tank environment where the "leisure" of research is understood, expected, and appreciated? That's because research isn't a linear straight-forward process. It involves many lateral moves and retreats which may be seen by some as failure… but for you, it may be earning perspective from different points that make up the issue-at-hand's "circle of illumination."

If your next goal is to be a researcher, find a place to work which fosters its environment. That way, you fit in with the company's or institution's culture.

Third:

Hard choices are often necessary if you want to get ahead. These choices which make things right for you may hurt the feelings of others; however, they're not the ones who have to live your life. You do. Make the choices right for you. This may mean that you hurt the feelings of the professor (or professors) for whom you do research. If you determine that their line of research isn't for you, it's better to be upfront and honest about it as soon as you realize it isn't your cup of tea. It happens all the time so while they might be disappointed by their investment in your potential towards their research, it's a harsh reality that not every graduate student will go on for their doctorate degree.

The last thing you want is to offer half-hearted efforts because they tend to lead to tepid results which disappoint. That's not fair to you, and it isn't fair to your professor.

Only you know what's best for you. Don't be afraid to honor and defend that quest.

Fourth:

During my first year in engineering graduate school, there was one course that stands out as being the most grueling class I ever attended. As I later learned, the homework questions in that class had never changed in the decades it had been taught… so the answers were floating around campus with a wide variety of sources. In many cases, one problem would take ten pages of math so difficult it made one's head spin.

Apparently, the class was a rite-of-passage because there was an unspoken expectation that the students in attendance needed to network with other, older graduate students who had taken the course in order to get the solutions to the homework.

As I learned, they didn't expect you to know how to do the math; they expected you to make friends so you could pass the class... which was graded on a curve... purposely.

The fact of the matter was that you were to learn that engineering a solution to a difficult problem is rarely done all by yourself. You need help from your friends, and that particular university expected you to make friends and peers... out of sheer necessity.

Fifth:
Take the time to express your gratitude and appreciation for what your research advisor contributed to your personal growth. It may require going above and beyond the call of duty; yet, gratitude is a gift we can give. Also, it fosters a sense of networking and trust so if you need to call on that professor's assistance years from now, you can consider them as an ally who can and will lobby on your behalf.

In my case, I graduated with my master's degree in 1992. After three years of working for a corporation, I sought out my research advisor and had breakfast with him at a local restaurant a few times in 1995 and 1996. Then, it wasn't until 2014 that I saw him again. By then, I had the perspective of really knowing what degrees of impact he imparted, and I was fortunate to communicate those nuances of appreciation with him... 20 years later.

It was a moment that enabled me to provide him with the context for why I made the decisions I did while I was a research assistant of his. At the time when I was still his student, I really didn't know or understand all the underlying reasons for my actions. Years later, I did understand, and it was an honor to share with him why I was so grateful that he took me under his wing. It was an opportunity to make another person feel good about their decision to be a professor and to help young students mature. Teachers need that from time-to-time. They need to know they made a difference.

Take the time to let them know their mentorship mattered. Explain how and why.

Not only will they appreciate it, but someday, so will you.

It's in giving back that we truly do receive.

Lecture Ideas for the Students in Your Classroom:

The importance of "publish or perish" in major universities should be communicated. It's the essence of why professors take on students to be their research assistants.

A leader delegates. A leader knows they can't do it all. In many cases, the professor sets his research assistants about the task of learning the details and communicating these findings for publication in a technical journal. The professor then often has the task of proofreading their combined presentation and tweaking the details with their nuanced understanding which has come with experience as a professor.

Some Homework Suggestions for this Week:

Have each student make a copy of an article in a technical journal and summarize what they learned from reading its research and findings.

My Ruminations about this Stage in my Life:

Zero level ideas and random walk theory were my two "take withs" from my days as a research assistant. The more important of those two turned out to be writing my master's thesis on random walk theory in association with tracking relativistic electrons flowing down the Earth's magnetic field in an attempt to create artificial lightning from space. Don't let yourself get bogged down by all those words though. Really, what was truly important was the concept of "random walk theory"... which is otherwise known as "The Monte Carlo Method." What it does is analyze a system – statistically – over many case histories… so you can discern an overall trend over time.

Said in an alternate way, it's like the movie *"Groundhog Day."* Bill Murray's character must learn that particular day's trends by living it over and over again… until he finally learns to do everything right… in order to reach the next day and get the girl.

This Week's "Take-With" for the Professor:

Consider taking what you learn from your students during this term of teaching "From Student to Engineer" and publishing your findings in some sort of educational journal.

Like research, this course is one of exploration. It's about what it means to be an engineer. It's about dispelling myths and incorrect concepts in order to arrive at an understanding of the value that comes with becoming an engineer. The more we share what works and what doesn't; the better off students will be at making wise choices.

This Week's "Take-With" for the Student:

Research is a quest into areas of the unknown and to come back with an understanding which needs to be shared with others in order to advance science into usable technology which makes life easier for future generations due to your diligence in areas previously foreign and misunderstood. It's a journey into the frontiers of knowledge with the pioneer's spirit where you forge a path and leave a trail for others to follow.

Moral of this Week's "Story"

Learning how to learn something new is a trait which will serve you well throughout all stages of your life. Once you "learn how to learn," it's like riding a bike… after you've done it, you never forget "how." Research is the art of learning the "how."

Curiosity is its core trait.

Curiosity helps create new scientific horizons. It is a characteristic which fuels daily passionate pursuits. Nurture its childlike fascination with what is not known.

If you do, you'll find the key to forging new frontiers which point out ways to a path that arrives at the fountain of your youth.

It makes your drive-to-thrive stay-alive all the days of your life.

Week 6:
To Get a Master's or a PhD? Post-Doctorate?

Some Brief Brainstorming:

There's no such thing as a sure thing.

The more you learn translates into learning how much there is to learn. When you get to this stage, almost everyone will be a prodigy or a genius.

Finding your niche and exploring it with all your gusto and all your might is crucial.

The Humor in this Stage:

Just as there a published book by Thorstein Veblen titled "*The Theory of the Leisure Class*," Robert Pirsig postulated that there should also be an accompanying "Leisure of the Theory Class." Pirsig is the author made famous by his book "*Zen and the Art of Motorcycle Maintenance*." Mention that "***Leisure of the Theory Class***" to a department head's personal assistant and you'll probably be able to get a laugh or two out of them.

Student Goal(s) for this Week:

Realize that grades matter when it comes to getting into the grad school of your choice. They also matter when it comes time for getting selected to be offered a teaching assistantship or a research assistantship… which pays your tuition, provides a salary to live on, and also gives you medical and dental benefits. These things matter when it comes time to graduate and start paying off your student loans. The fewer loans you have; the lower student debt you have to pay off. That equates with greater freedom later on in life. If you have a lot of student debt, you have to work to pay them off. That can shackle you to a job you may not like for years after your graduation. Consider it.

The reason we get good grades today… is tomorrow.

Professor Goal(s) for this Week:

Revisit the realities of "publish or perish" for a university professor and communicate this expectation to your students if they end up getting their doctorate in an educational research institution and then going on to teach in one as well.

If you're a professor, describe your personal educational path.

If you got your PhD, explain what it did for you.

Also, communicate the requirements to earn a doctorate: a written dissertation and an oral examination where you must present your findings to the public… as well as to a panel of professors who will be the ones to actually confer the PhD status afterwards via the appropriate channels within the university which enable the degree to be granted and your graduation to ensue shortly thereafter.

The Benefits of this Stage in the Engineer's Journey:

Learn how to learn

If you have a personality that enjoys identifying with strong fundamental research, then getting a doctorate and staying for post-doctorate work may be right up your alley.

In essence, doctoral and post-doctoral work ensures that you become one of the world's foremost experts in a very narrow specialization of science studies.

The Downside of this Stage in the Engineer's Journey:

Lots of studying material you don't understand.

When you look at publications in technical journals, a lot of it is like reading a foreign language, and if you are to develop into an expert in this area, you must learn that language and become proficient in producing it because it's a "publish or perish" world out there for advanced graduate students and professors.

A PhD in physics or engineering doesn't necessarily ensure a stellar earnings trajectory. Advanced degrees don't always equate with getting the job of your dreams.

Another downside of this stage is that by being one of the world's foremost experts in a very small sliver of science, you may only have a handful of peers who really and truly understand you and what you do and what you're developing into a dissertation.

If You Don't Know Your Destination, How Will You Know When You've Arrived?

Risk of burnout is a real reality at every step of your life. This is especially the case in graduate school. There, you may find yourself giddy to get into life's saddle which often feels deferred when you stay in graduate school. However, if teaching at the university level is your goal, then getting a doctorate should be a specific destination.

Top 5 Things to Do at the Start of this Phase of Your Life:

First:
Envision your perception like headlights that only see so far down the road. Where do you want this road of advanced studies to take you? Is it the right road for you? That's because graduate research often seems liking driving a lot of darkened roads where it's only your ingenuity and learning perseverance that light its path.

Second:
Consider taking a class in grant proposal writing. A lot of research institutions offer these sorts of classes in their School of Social Work. Find one. Take it.

In graduate school, knowing what it takes to be rewarded with grant money is essential for continued research. Learn this skill. It will serve you well all your life.

Third:
Make sure your interests are well-served. Ask yourself, "Where do I want to grow?"

Fourth:
A great player respects a winning streak.

If you're doing well because of the habits and superstitions you're obeying, respect the streak you're on and keep on doing what you're doing. Sometimes, this runs contrary to what people may tell you is correct, but if you're doing well, the worst thing you can do is think too much. A lot of times, you must not think… and just "do"… especially when you're "winning."

Lucky streaks don't happen often.
Respect them.

Comedian extraordinaire, Jerry Seinfeld, is said to have referred to it by saying, "Just don't break the chain." As the story goes, he would have a large calendar on the wall with all of the days of the year on it. Whenever he'd write new material, he'd put a large "X" on that day. To him, it represented a day of accomplished achievement. The more days in a row that he wrote; the more X's appeared so that daily progress formed a chain of X's on the calendar. If he had a day he didn't create new material, there'd be an empty space which broke the chain. He'd remind himself, "Just don't break the chain."

Consistent accomplishment develops a momentum. Keep it going. Once that momentum stalls, it's hard to break free from the inertia it takes to regain it. Respect the streak.

Fifth:
When you're in the PhD or post-doc environment, all of your peers are prodigies and geniuses of their own field of study and

specialization. That's where it's up to you to discover, "What's my thing?"

In other words, "What separates me from the rest?"

Top 5 Things to Consider at the End of this Phase in the Engineer's Journey:

First:
If you want a full-time job as a tenured professor in an academic research institution, you must get a PhD. It's a job requirement if you want to teach at that level.

Second:
An advanced degree matters if you decide to detour away from your line of study to pursue other areas of life.

Here's a joke from the film *"School of Rock"* – "Those who don't 'do' teach. Those who don't teach, teach gym."

Third:
Analyze misconceptions.
Meet with your advisor.
Figure things out.

Have an idea of what you want to get out of this stage.

After all, you'll be at it for quite a while.

One thing you'll want to do along the journey of the scientist and the engineer is to revisit old theories and ask what assumptions were made which were wrong. This way, we arrive at new discoveries and new theories which work better. It's a lifelong process.

Fourth:
Decide whether you want to be a scientist or an engineer. There is a difference. There's a gradual overlapping between the two where physics turns to applied physics which morphs into applied

science… and then eventually becomes engineering. Should you decide to steer your ship towards the port of the PhD, consider where you'd see yourself in that spectrum.

Fifth:
Another thing to consider here is whether or not you'd be a good fit for government work. If you do get a job doing work for the government, there's a good chance you'll have to get a security clearance. Usually, this is something your company will pay for.

This means that you'll be working with intellectual property which is "sensitive" and "proprietary" so don't be surprised if you and your work face a lot of scrutiny.

Also, the application for a security clearance is thorough, involved, and very long. It may require a lot of research of your past to get things such as all your past addresses. Complete and utter honesty is expected… even if it means an examination of some parts of your past you're not so proud of… and then letting the chips fall where they may.

The reason I mention this is that the sooner you realize that you may someday get a security clearance; the sooner you'll make sure you start living a life whose every aspect may someday be scrutinized for future success in your career. It's better to be aware of it in order to be proactive rather than reactive later on.

Lecture Ideas for the Students in Your Classroom:

Emphasize the long-range thinking that exploration and research requires for innovations and their inventions to thrive.

The further you can get people to look ahead; the more their graduate-level explorations will pay off. Exploratory activities are long term endeavors. Are you up to it?

This is especially important if those post-graduate level students have to justify their decisions to continue onward with schooling for years and years to their parents. Parents may rationalize with pragmatic attitudes that can make it awkward to continue on the road of being what they might consider as "a perpetual student."

However, if such studies enhance your professional identity amongst your academic peers, such decisions are justified. That's because a lot of research think-tanks tend to recruit their intellectual cronies so they can facilitate groundbreaking collaboration amongst trusted colleagues.

Some Homework Suggestions for this Week:

Ask each of your students to find someone's PhD dissertation, pull it off of the shelf, open it up, and explore that particular person's passion in order to get their doctorate.

Then, have them jot down their immediate ruminations over witnessing the culmination of someone's blood, sweat, and tears for 5-6 years.

While they can probably do this online, it'd be better if they visited a university and wandered into "the stacks" of a graduate library, blew the dust off the cover of that particular dissertation, and gazed at a bit of history… gazing at one person's past and pondering how that dissertation made a difference in that particular person's life.

Another idea is to have your students explain why consistency doesn't necessary equate with it being a valid proof. An ancillary question is to ask them why they shouldn't ever trust their science 100%.

My Ruminations about this Stage in my Life:

When I began writing this book, I lived in the area where I went to grad school and marveled at September student-move-in congestion. I also reminisced over all the old storefronts that have now changed into new stores. Some storefronts have had a high rate of turnover and gone through many different iterations of names and ownership. It's a reality that many small businesses don't make it in the long run. High overhead for brick-and-mortar stores is a reality… so is the fact that university expansion often causes buyout offers for small business owners. Good universities do that. They expand.

That doesn't mean that you need to align yourself with its expansion plans though. Universities are always recruiting. Doctoral candidates are their target. Even in undergrad, you may feel like you're being recruited for grad school and lobbied to get your PhD. It's their goal because you stay at their school the longest amount of time, and your research results may eventually enhance their elite reputation and prestige.

In my case, I considered the PhD path an awful lot because I wanted to do space-related science that would have impact and ramifications for NASA's healthy progression. When I teamed up with my research advisor, the reason I targeted him is that he had an experiment going up on the space shuttle, and if I stayed on to get my PhD, I'd have spent time at NASA's flight control and mission central to receive my data which would eventually become the premise of and basis for my doctoral dissertation. It was called the Shuttle Electrodynamic Tether System (or SETS for short). Essentially, it was a satellite on the end of a very long rope which would be unreeled out into space and swept through the Earth's ionosphere. To me, it seemed like really cool science to do.

After much discernment though, I decided the PhD path wasn't for me. Part of that decision came from outside sources… such as falling in love and wanting to earn a living that could support that love comfortably and as soon as possible… so I decided to wrap up my graduate studies and exit the university setting with a master's degree.

In hindsight, I found I'd made the correct decision because both times the experiment went up on the space shuttle, the tether snapped… sending the satellite at its end drifting off into space. The funny coincidence about it was that each time it failed, I "just happened" to be on vacation with friends in locations we considered to be paradise. The last time it failed though, the TV news showed a shot of the satellite and its attached length of snapped tether drifting away into space. When the shot returned to the gorgeous female TV anchor, she quipped, "Wow, it looks like a giant bungee cord."

I was flabbergasted. My whole potential PhD just got summed up in a simple sound bite.

The funny thing now is that when I've revisited my old professors and advisors who were associated with the SETS project, when I mention that line, we all laugh. We all get it.

This Week's "Take-With" for the Professor:

While I know I repeat myself from time to time in this textbook, I do so consciously for a reason. I do so in order to form a sense of "muscle memory" in the mind of the student. In essence, it's the reason hypnosis works... repeated suggestion.

Like at the aerospace corporation I used to work within, we used to have a job called "an expeditor." It was the person who nagged in order to get things done and completed. A fellow professor friend joked that his ex-wife would be happy to do that job for free.

However, in all seriousness, at that corporation, one of our managers once said, "If I haven't heard from someone 5-6 times about something, then I know it isn't really important." In other words, the squeaky wheel gets the grease.

It's the essence to my method of madness for why I use "repeatability."

This Week's "Take-With" for the Student:

Staying on the same page is an important issue that I learned from a NASA astronaut. He explained that he reverse-engineered many of the space shuttle's systems to gain an absolute understanding of how they worked, and he kept his notes in a journal. When his superiors at NASA found out about these journals, they made absolutely certain that he make copies of every single page. They stressed that it was of vital importance that they have the same set of notes on the ground as he had up in space… so if something went wrong, they'd all be on the same page when trying to troubleshoot should anything go wrong while up in orbit. His participation in Apollo 13's safe return was a case-in-point.

Moral of this Week's "Story"

Life is made up of many tides. It is often like a sinusoid. The issue isn't about maximizing the pinnacles. Instead, it is all about minimizing the troughs. When you realize that you are in a valley, it takes some soul-searching and some much-needed changes that cause you to dig down deep and remove that "shard of glass" that tends to exist inside each person's psyche and soul. It turns a "zero" into a "hero."

One can be too optimistic. The problem with this is that it is like dating a movie star. At first it seems great and is all that you ever dreamed of, but sooner or later, they will probably break your heart. This optimism tends to reside the most strongly when you are living in a boom time… or a bull market. It seems as if nothing can go wrong. Usually, that can last for only so long. The goal is to prepare for it. This doesn't mean dodging the storms. Instead, it means learning to dance in the rain.

Week 7:
Walled Garden vs. Real World

Some Brief Brainstorming:

Caterpillar to butterfly emergence takes time. When do you want to fly out in the real world? Or do you want to stay stuck in the cocoon as a perpetual student?

Once a dragonfly emerges from the water, it can never go back.

The Humor in this Stage:

Engineers often work in a world of black and white… which is why I was so astounded when painting a mural, sketching its outline in black and white, and then being told, "I can't wait to see what you do when you add color."

Student Goal(s) for this Week:

You need a PhD to stay inside the walled garden of the university.

What do you want to do when you grow up?

Professor Goal(s) for this Week:

Convey Benjamin Franklin's quote: "Well done is better than well said."

Doing what you're good at isn't the same as doing what you love.

The Benefits of this Stage in the Engineer's Journey:

You often only get one chance in life
Study what you want to study.
Exploration and freedom.
Find your passions.

Explore them.
Transition.
Refine.

The Downside of this Stage in the Engineer's Journey:

Life has a way of "raising the ante." There will be a fear you'll have to face upon finally finishing your schooling. It is a crossroads you'll eventually arrive at. The thing about being a student for so long is that you become comfortable with the environment and its expectations. Departing the university setting is daunting and even a little bit scary. We fear what we do not know or do not understand.

The one thing you don't want to do is to lie to yourself about how you feel regarding a "real job" out there in the "real world" after graduation.

It's a slippery slope to lie and then to lie again to cover up your lie. This turns a lie into many lies. As my mom would often echo, "The dirt always comes out in the wash."

Be honest with yourself. Some students become "perpetual students" who never want to leave the university's nest. The problem with this is that at some point, those people need to acquire possessions such as a home, a nice car, and so on. Also, the perception from people in society is often associated with the position you hold in some job.

However, as one professor of mine in Beverly Hills once said at the end of one of his lectures, "The goal isn't always to get a *good job*. Instead, find value in doing a good job of *being Real*."

The goal is really to transition into a form of self-sufficiency.

If You Don't Know Your Destination, How Will You Know When You've Arrived?

Fermentation and distillation "Transformations" matter. Fermentation is a first order transformation. Distillation is a second order transformation. Using this analogy, undergrad ferments you into an engineer. Grad school distills your engineering essence.

There is a time when the baby bird must leave the nest and learn to fly.

A university's campus, to me, is a walled garden. It contains different expectations and perceptions than out in the real world where one must have a job to be perceived as successful. The walled garden enables exploration to flourish. In the real world, there are deadlines to meet, work to do, and bills to pay. However, in the real world, you are usually compensated with more money than within the walled garden. If making a living out in the real world doesn't appeal to you as much, consider getting a professor position on campus in order to stay within the walled garden environment. This usually requires a PhD if you're going get a professorship at a research institution. That helps you tune into your overall destination during your days of enjoying campus life.

Top 5 Things to Do at the Start of this Phase of Your Life:

First:
Realize that the path to success almost always travels through failure first.

Second:
Realize that you're going to go through lonely stages in your life. Music is a character who is always willing to be your constant companion during these lonely valleys which you must cross – yourself. Music will always keep you company. Allow it to be your friend. Even better, learn to contribute a few notes and stanzas in its midst.

Another thing to consider? Join a band. Play music with others. It's one of the most important non-verbal communications you'll ever employ. Music isn't about playing a song. It's about communicating with fellow players in a way that present a song.

Better yet, you don't need an instrument to practice. If you can't sing it, you can't play it. Often, my steering wheel gets more practice while I'm driving than my actual drums.

If this is a time of freedom and exploration, add Music as a vital ingredient. Not only does it make you a better-rounded person, it may open avenues you'll never imagine as an engineer alone. It did for me… and I never saw it coming until after it arrived.

Third:
Miles Davis said, "Don't play what's there. Play what's not there."

Fourth:
Make sure to take jobs where you have the chance to look back at the end of the day and say, "Yeah, that's good." It's the essence of earning a satisfying career.

Be able to take a step back and say, "Good." When you do what you know you should do, you give the gift of truly living.

Fifth:
Try to find satisfaction in every step of your journey. It leads to overall happiness.

Also, "humility jobs" shouldn't be looked down on as being "beneath you." They add to your arsenal and provide communication lessons with people from all walks of life… not just the ones that require a college degree. In order to help make ends meet in college, you may have your fair share of these humility jobs. Embrace them. The attitude with which you do any job directly correlates with your overall enjoyment every single day.

This time in your life enables you to savor the thrill to search for truth. It's also a time when you also should search for your own truths too... not just overall truth. Once you enter the real world and get "a real job," there's almost an unspoken truth in career corridors that if you look like you're having too much fun, you become suspect.

Ask yourself when you're willing to trade the ivory tower idealism that thrives within the walled gardens of almost any college campus for the nitty-gritty, down-and-dirty harsh reality of products made with "planned obsolescence" so consumerism continues to enable the need for the company's latest-and-greatest product-purchase. That's how companies survive and succeed. After all, the best customer is a repeat customer.

Averse reactions to risk are a human reality. People are afraid of what they don't understand. This includes breaking through that incubating glass bubble within a college campus in order to begin your money-earning career in "the real world." It's similar to the fear a child faces when going from elementary school to junior high school. Then the same sort of fear must be faced when going from junior high school into the bigger high school building where facing puberty while learning and growing is an awkward reality. After that, another fear occurs when searching for the correct college to apply to with hopes of getting accepted, and having to learn to live on your own in many cases while the level of study intensifies. Notice though that all of these former fears were all about going from one school to another. At some time, those "walled gardens" (aka – schools) graduate you to the next level where "the real world" beckons with its "rat race" that children see as they witness the trials and tribulations of their parents. It's a scary thing.

Top 5 Things to Consider at the End of this Phase in the Engineer's Journey:

First:
How many times have you walked into another room and forgotten why you came into the room in the first place? It's a common occurrence.

In both the real world and within a walled garden, there are many doorframes and portals that we pass through which rob us of the memory for why we passed through its threshold. It's been tested, studied, and (somewhat) proven that the "doorframe portal" matters when it comes to memory. That's because the moment you enter another room, your surroundings immediately change, and the senses become overloaded with new views, smells, sounds, etc. Thus, one way to remove this from being an issue is to consciously repeat what your purpose is when entering a new room. Actors do this when they're remembering their lines to a scene they're about to deliver. They repeat their lines over and over to themselves while waiting for the crew to set up for the next "take" when the director calls "Action!" It works. It keeps your priority on top-of-consciousness so a great idea doesn't get lost within your memories.

This is why it's so crucial to jot down new ideas as soon as they arrive in the forefront of your memory. Once they get shifted "down the stack" and its "order of importance," a good idea can get lost in the shifting of the mind's internal kaleidoscopic workings.

Second:
Regardless of your position or lot in life, be one of those persons who stops and asks for directions when you are lost. Studies show that successful students often reach out for additional resources when they are unsure of how to approach the next task.

Third:
Remember that some people, after years of schooling and real-life experience will long for the simplicity they sense with daily accomplishment. In other words, as my senior scientist mentor once told me, "The happiest people I know are creative daily."

Fourth:
Battles are won before they are fought.

Reconnaissance matters. It provides priceless perspective.

There's a lot about life you'll never learn inside the "walls" of a university campus.

Fifth:
Street smarts beat book smarts in a lot of different avenues Life presents.

The power of being "good people" matters. It's like being able to look someone in the eye… and firmly shake their hand… no matter what walk of life they come from.

Even street people can teach you the world. They can teach you how to build a house and sail a boat… by doing it… rather than a professor assigning you to read a book on how to do it. While they may be money-poor, they are time-rich with stories to share.

Lecture Ideas for the Students in Your Classroom:

One topic that should be discussed at some point is "having children." It's one of the reasons a lot of people truly get into "the real world" because now, they have mouths to feed and children to support besides themselves. As Bill Murray said in the movie "*Lost in Translation*," "It gets complicated when you have kids."

Some Homework Suggestions for this Week:

Describe and discuss the differences between wildflowers and those which are planted in a garden. Then touch upon what happens if the gardener stops tending their garden.

Then apply it to the image of an abandoned building which is now overgrown with weeds which grow between the cracks in the cement and pavement.

My Ruminations about this Stage in my Life:

At one stage of walking out of the "walled garden" and into the "real world," I remember looking in my rearview mirror as I drove away from where I lived during a summer of research that resulted in my first published work in an engineering technical journal. While departing, I found myself listening to the solemn Don Henley song, "*The End of the Innocence.*"

Whenever I hear that song, I remember the rearview-mirror point-of-view of my foreign-exchange-student-friends waving "goodbye"... for the last time I'd ever see them.

It's the kind of moment that occurs when the turning of the page actually reveals the end of a chapter. It's okay if tears arrive. They did with me. Somehow, they cleanse.

Looking back, the most important thing about that time in my life is that I forced myself to move to a location where I didn't know anyone and had to live on my own. There was a fear to face in doing so. It's a going "*into the New*" kind of feeling. It is important to face this, sooner than later, so you possess the confidence that you did it. Knowing that provides a confidence that if you have to move to another new city, it'll be okay. You will survive. You will make new friends. You'll have a clean slate to work with that is free from old memories and is a blank canvas with which you make a new masterpiece.

This Week's "Take-With" for the Professor:

People are afraid of what they do not understand... often, this fear is based around what a person does not understand about him or her self. "Transmutation" is the act of becoming someone new. It's a transformation... a lens-like... a jewel-like... a prism-like transformation which occurs when one continues to explore... and refine.

Transmutation is also the alchemical conversion of base metals into precious metals... such as silver or gold. It is a change from one nature to another.

Going from the walled garden to the real world is a real-life transmutation.

Encourage your students go from "zero" to "hero."

This Week's "Take-With" for the Student:

A lot of millennials would rather earn $40,000 per year at a job they love than $100,000 at one they hate. Regardless of what you do, just make sure you do it right and do it with enthusiasm. Life is too short to have a ho-hum attitude about what you do.

Moral of this Week's "Story"

Before you actually go out and work in "the real world," a lot more people in their twenties take a lot longer "finding themselves"… such as postponing marriage until they find out what they really want and are looking for because they've dipped their toes in a lot of different waters enough to know what matters to them most.

Transitioning into innovation's arenas, keep in mind that entrepreneurs are the kind who commit to company growth in ways where the weekly paycheck isn't their top priority.

<u>Week 8</u>:
A Student's Guide to Getting a Good Job

Some Brief Brainstorming:

Nobody's perfect.

Everyone's got something… an Achilles Heel.

Is it something your employer can live with?

If you can afford it, consider being an intern.

Being an intern gets you inside of the ballpark on the field which you hope to play and become a valuable player.

The Humor in this Stage:

Be flexible enough to laugh at your ignorance so you take on the open-minded mindset which enables you to learn new tricks that make you more valuable in the long run.

As a matter of fact, Jonathan Lockwood Huie is quoted as saying, "Life's burdens are lighter when I laugh at myself."

Student Goal(s) for this Week:

Look in a Sunday paper's want ads for engineering.

Start to realize what requirements a lot of companies ask for.

What are the skills they want? Pay attention to how much experience they require for positions that look attractive to you.

In many cases, a lot of jobs want at least 5 years of experience… which is why it pays to stay in your career at least that long before thinking about moving on.

Professor Goal(s) for this Week:

You can tell a lot by looking 'em in the eye. That's what a lot of hiring managers must ascertain as they interview you.

Teach your students to alleviate the doubts in the mind of any hiring manager by maintaining an air of flexibility to help as needed.

Also, instruct them to target a company if they really want to work for that company. It's a "rifle" approach rather than a "shotgun" approach to getting the job they desire. It means doing their research on who works there so they can try to network and connect with their employees. This can lead to a valuable referral from within.

The Benefits of this Stage in the Engineer's Journey:

Networking starts in college. If you're active in engineering societies and attend the meetings at the professional level, it can lead to grooming contacts that later lead to summer internships. This happened for me when I got an engineering intern's job during the summer between sophomore and junior year.

The Downside of this Stage in the Engineer's Journey:

Sometimes, affording an apartment right after graduating from school isn't easy. The further you live from your job; the more stress you add to your life due to the commute. Also, another reality is that some housing locations require roommates to pay the rent.

If You Don't Know Your Destination, How Will You Know When You've Arrived?

After you see the area, the company, and the actual location within the company you might be working for, you have to ask yourself if you can envision yourself living and working in this culture, these surroundings, and this environment.

Top 5 Things to Do at the Start of this Phase of Your Life:

First:
Take ownership. Take responsibility. Show integrity.
Then let the truth/consequences fall where they may.
The truth is always interesting, and it sets you free.

Second:
If you make a mistake, know that it's a step towards a solution.
The goal from it is to learn how not to repeat the same mistake
again.

Remember, goals and sunshine help you grow.

Third:
Discipline and freedom tend to go hand-in-hand. This especially
applies to your spending… which leads to your bottom-line
finances. Some say, "The past doesn't matter. It's the future that
counts." While this is true, the debts of your past do accumulate
interest which reduces your consistent buying power until you
pay off that past debt. Credit scores do matter. Stellar credit
ratings make life a lot easier.

Fourth:
Clear a path from daily "momentum busters" early on so a
streamlined process of consistent efficiency is enabled.

Fifth:
Know your resume. Don't represent "vapor ware" on your
resume.

Body language matters to hiring managers… so does a good
handshake.

Consider the commute. Shorter and easier commutes provide
greater peace of mind, and it also makes it easier to put in
overtime (greater pay rate for OT). Longer commutes equate
with vehicle depreciation, maintenance, wear & tear, police,

speed traps, greater probability for accidents, weather slow-downs, traffic tie-ups, bottlenecks, tricky travel conditions, landfills, and those orange construction-zone "anxiety cones."

Groom your professional page on business-minded social media websites (such as LinkedIn) and view the page as how an employer would see it.

Top 5 Things to Consider at the End of this Phase in the Engineer's Journey:

First:
The Buddhist saying applies...
"Before success... chop wood, haul water.
After success... chop wood, haul water."

Second:
Engineers can tend to be some OCD kind of folks. Obsessive-compulsive-disorder is easily confused with the thoroughness required for engineering excellence to thrive. In engineering, you must close every circuit for the electricity to flow. This borderline obsessive behavior ensures that questions get answers and that problems are solved with a dedicated follow-through that ensures the result is as accurate as possible.

This attention to detail is crucial in some fields. Engineering is a precise case-in-point. It's very methodical. It labors to a sense of closure so you can either rule out that particular method or embrace it... which is the essence of the scientific method.

This OCD-like behavior may make you seem silly to some; however, others will realize and understand that this precision is an essential ingredient in the recipe for success to be achieved at every stage that your life may pass through as your future unfolds.

It's the essence of the "alpha" personality... extreme thoroughness, precise attention to detail, the determination to make sure every nuance is covered… for the best decisions to be arrived at… with the most complete set of data achievable.

Third:
Realize that a cover letter is an essential ingredient if you're applying for a job. Skipping one can get you skipped over by your hoped-for employer. A cover letter gives a chance to show you've done your homework about the company at hand. It shows your degree of communication skill. It personalizes you in relation to your resume. It captures the quintessence of your resume in reference to the company you hope to impress too.

Fourth:
When a company lists the requirements they desire, understand that it's really just a *wish list*, and that you don't have to have every item in the list within your arsenal's toolbox.

Fifth:
Housing affordability issues (such as in San Francisco and Los Angeles) are prime concerns whether to take a job for the salary you're offered. For example, in 2000, I was offered an engineering job near San Jose with a $55,000 per year salary. That's about a thousand dollars a week… before taxes and other deductions. When I scouted locations to live near the plant I'd report to, single bedroom apartments averaged around $1500 per month. Thus, based on those facts alone, I ascertained that the job offer's salary wasn't enough to enable a comfortable life in an apartment free of roommates. From the outside though, I was perceived as crazy for turning down "a good job" offer by my friends and family in Detroit (where a salary of $55,000 a year offers a bigger bang-for-the-buck).

It's often hard to negotiate a higher salary as an outsider from the company. You either take their offer or you leave it. This is where it's nice to have a few job offers to choose from. In my case, once I graduated with my master's degree in 1992, I got offers in Los Angeles and Dallas. Since my sister was attending UCLA at the time, the fact of having some family near my first "real job" made it easier to decide on working in LA. I sometimes speculate what my life would have turned out like if I'd taken the Dallas offer, but I can't go back and wonder "what if" with all the books I've now published and with all the wisdom life's experiences have provided.

The whole point is that you need to take some time to discern if the job offer you receive is really the one you want and is where you want to live. This is where doing some reconnaissance of the area (and its prices) is crucial when you have your job interview.

Lecture Ideas for the Students in Your Classroom:

Here's an in-class assignment. The students must participate for true understanding to be obtained. It's an exercise in learning process efficiency... which relates to multi-tasking. (Author's Note: I mention this lesson next chapter too. To me, it's *that* important.) Explore the "index card assignment." Here's the task. Place two competing teams of equally numbered people across the table from each other. Give the first person five index cards. Have them sign each index card five times... using three different methods.

The first method is for each person to sign the card five times before they pass that card to their adjacent teammate. Record the amount of time it takes to complete the task.

The second method is for each person to sign the card three times before they pass it to their adjacent teammate. Then the person at the end must pass the card signed three times to the first person again... to sign two more times... for a total of five. That person passes the card to their adjacent teammate again to sign two more times so each index card is signed five times by each teammate. Again, record the time it takes to complete this second process. It should be shorter. It's a more streamlined process.

The third method is for each person to sign the index card once before passing it and returning it to the next teammate. The five index cards still need to get signed five times. This time, each teammate isn't waiting as long before signing. The cards are being passed with minimal waiting. This process is the shortest.

The lesson learned from this exercise is examining the efficiency of three different processes. The one that takes the longest has the most amounts of teammates waiting while others complete their task. The process that takes the shortest is the one which has the least amount of down-time between teammate tasks.

Some Homework Suggestions for this Week:

Write a cover letter for a fictitious firm you'd like to work for.

These days, many jobs you apply for now require a cover letter.

This is why a course in technical communications is essential.

So make sure your students are able to write well.

My Ruminations about this Stage in my Life:

"When you are ready to manifest your destiny, it will come to you." That was told to me in Beverly Hills, California in the summer of 1998. While it can be frustrating when that "destiny" of yours doesn't seem to be manifesting, when it does, it usually does so in such an elegant way that it leaves you stunned in awe and speechless with amazement.

This has been the case in my life. Looking back, I've noticed (as have hiring managers) that I don't have a sense of longevity with any of the jobs I've worked. However, I do have a sense of breadth in the variety of positions I've participated within. It's the primary reason I'm able to write this textbook with a sense of tangible authority… because I've lived through each one or been exposed to its nuances via my closest friends and family in order to understand it first-hand. For example, while I didn't get a PhD, my first true love did, and she'd share her concerns and cares during our relationship while she earned her doctorate. Also, in grad school, you talk to a ton of PhD candidates.

They say that you write what you know. Thus, this textbook is one which, in hindsight, I can see that my job-hopping has prepared me for in order for these lessons to manifest.

It also alludes to the fact that I've had to search for numerous jobs. Honestly, the easiest time I had getting a good job happened when I was about to graduate from grad school with my master's degree. At my university, at the time, their engineering placement office resided in a building with a central lobby, a few adjacent offices, and then a long hallway filled with small rooms containing a table and two opposing chairs where the job interviews would take place. They had a book of companies that were coming to campus, their employment needs, and the times and dates with which you could schedule an interview. These 15-minute on-campus interviews led to me being flown to Providence, Rhode Island… to Tampa, Florida… to Dallas, Texas… to San Diego, California… and finally, to Los Angeles, California. Those five on-location 2nd interviews were all-day-long tours and interviews with numerous managers and employees who would then decide if I was a candidate they'd extend an offer to.

Just make sure to let your professors know ahead of time about these on-site job interviews so you maintain accountability and responsibility for your class work.

It's a neat feeling. The interested company tends to send you plane tickets via Federal Express one-day-mail. You get put up in a nice hotel, and sometimes they'll have you spend a few days in town to see how that city feels for you and whether you'd want to live there. Sometimes it comes with a car rental, and sometimes it has a designated employee pick you up and drop you off at the hotel, the job site, and the airport.

My advice is to explore the area as much as you can and the prices you'll pay for rent. Nowadays, they also have relative "buying power" ratings of one dollar. Thus, a $100,000 per year salary in New York City doesn't stretch nearly as far as the same salary in Detroit.

An example of this is that a company's headquarters that I interviewed with in 2001 was in Detroit, and there, they asked me what my salary requirements were… a common question asked in first interviews… so do your research. I replied $55,000 per year. It was an average engineer's salary in Detroit at the time. However, the actual job which they wanted me to work was located near San Jose, California… where $55,000 per year doesn't go nearly as far due to the San Francisco Bay area's high cost-of-living. I didn't realize this in the initial interview at that company's headquarters, but I did realize it after the job tour they gave me near San Jose; therefore, I turned down their formal job offer. Also though, when I took the job tour, it dawned on me that the job was not for me. While it sounded great on paper, when I saw the actual surroundings and environment I'd be working within for at least 40 hours a week, I knew immediately that it didn't match my personality whatsoever. While a lot of friends and family scoffed at my decision, they're not the ones who took the tour or would have to work the job… so sometimes you have to stick to your guns and be choosy about what you feel is right for you.

You don't have to take every job that is offered to you. Some just aren't a good fit. There's nothing wrong with realizing that. However, it is better to find that out before you take a job than after you take the job. It's like finding out that you married the wrong person because you rushed the engagement period. That's because a job is like a marriage in many ways. It's better to be in love with what you do. You're going to be spending a lot of time doing it, so it's best if you actually like it a lot. That way, it tends not to seem like work… which is a good way to feel… especially at the end of a long day and after a long life. Life's too short not to enjoy what you are employed to do – daily.

The last thing you want to do is to live with regret.

Taking the right job helps alleviate this issue.

This Week's "Take-With" for the Professor:

Share this quote in your class. It's from Stanley Tucci in the fine fashion film *"The Devil Wears Prada"* – "That's what happens when you start doing well at work. When your whole life goes up in smoke? You're due for a promotion."

This Week's "Take-With" for the Student:

There's no such thing as a little mistake.

Also, all job-search advice you receive may not be good advice. It's up to you to discern what is true. Adrienne Rich is quoted as saying, "The unconscious wants truth. It ceases to speak to those who want something else than truth."

Moral of this Week's "Story"

Social Darwinism is a cold fact of life... which means "survival of the fittest."

Part of that survival is realizing that you have a brand. Each social media website you belong to presents and delivers a slice of your "brand." Overall, those pieces present an entire pie to potential hiring managers. Your presence on the internet delivers a vast amount of information about you. Try now to make the most of your internet perception.

Really, your top product... is you.

Professionalism starts now.

Week 9:
Engineering Realities in a Large Corporation

Some Brief Brainstorming:

In 1995, after three years of corporate engineering, a fellow engineer asked me, "Are you institutionalized?" He asked after his first viewing of "*The Shawshank Redemption.*"

Bad ideas to do science? When they are intellectually fallacious.

Discuss the streamlining of process performance.

The Humor in this Stage:

Laughing way too loud over lame jokes… stress relief measures. This is something you might encounter after too many hours of overtime at any corporation. Someone will tell a joke, and someone will laugh… causing a domino effect of infectious laughter to result.

Student Goal(s) for this Week:

Professionalism is in the details. Employ it at every stage you perform. The sooner this habit becomes second nature; the better you'll do in every duty you do.

Professor Goal(s) for this Week:

Encourage your students to learn from mistakes. Grow into getting along with all people. No judgment is a crucial component to practice because many managers in a corporation don't actually work with the hardware… whereas the technicians may be the ones who can contribute the best ideas on how to make things better because they actually work hands-on with the product that will ultimately be utilized by the consumer.

Also, teach them to tolerate frustration. At the start of a student's corporate career, rarely will they make tangible contributions until later on.

The Benefits of this Stage in the Engineer's Journey:

Learning how to pitch a big idea and organize occurs a lot in a corporation. The best way is to get it out of any one person's mind so that the rest of the team can analyze its strengths and weaknesses. A case in point occurred on a proposal for 15 satellites. Our project leader procured 3-4 adjoining rooms where each wall would be dedicated to presenting our data in order to prove our worthiness to win the contract. Thus, antenna analysis PowerPoint slides with its evolving performance iterations took up one wall... while propulsion took up another wall... and payload took up another... and so on.

The cool thing is that when a new design rose the bar on that particular system's performance, a new PowerPoint slide was generated and placed over the previous one. None of the prior slides were pitched out. That way, they could be compared like flipping through the pages (months) of a calendar at a glance. Also, by having the entire satellite constellation's system up on the walls of all these rooms, there were many sets of eyes focused on tweaking, editing, and reordering them into the best presentation possible. Truly, it was a team effort that benefited from having that "bullpen" organization as its approach as we designed the satellite system over two years.

The Downside of this Stage in the Engineer's Journey:

The Glassy-Eyed Look that comes from Information Overload. It's a lot like you feel as if you're taking a drink from a fire hydrant. Sometimes, jobs can grind the gears off of the bright and shiny cog you once were when you got hired fresh out of school. It's a sad reality that the daily grind can precipitate a nervous breakdown or a mid-life crisis, and none of us are immune to it. Some of it is caused with your sense of value within your job and how the company perceives your contributions. This goes hand-in-hand with your sense of job security. No one likes to go to work where they feel like they're walking on eggshells. You always want to have your heart in the job you do.

The bigger a company is; the better it is at eliminating innovation.

If You Don't Know Your Destination, How Will You Know When You've Arrived?

"Wanting" to work rather than "having" to work is essential for career happiness. Even in a corporation, you can be happy in doing what you feel you're perfect for. The way to maintain this happiness is by reducing outside sources of stress… such as living beyond your means in a way which forces you to work more than is healthy for you, your outlook, and your mental state of well-being. For example, one engineer I worked in a corporation with bought a Corvette and an expensive house in a nice neighborhood that was far away from where he'd do his job. It required he make a long commute and that he work overtime every single week in order to pay all of his bills.

Top 5 Things to Do at the Start of this Phase of Your Life:

First:
The trick isn't getting hired.
Often, it's staying happily employed.

Real passion can't be faked. If your bosses start to sense that you're not genuinely enthusiastic about the work you're doing, you might find yourself getting measly raises and perhaps even a pink-slip lay-off. As legendary Green Bay Packers head coach, Vince Lombardi, (loosely) said, "If you don't play with passion, you'll be fired with passion."

Authentic interest in the work at hand demonstrates commitment. A company needs that to thrive. It also fosters an active career that stays vibrant and alive.

Second:
An employer's trust in you is a crucial component. It's essential to groom their trust. That also means keeping quiet about some of your personal life. For example, if you happen to come in with a hangover, that doesn't mean you have to mention your festivities from the prior night of partying. Sometimes, less is more with what you communicate to others you work with. Reputation and character go hand-in-hand. If you develop a reputation for being untrustworthy, your stay at the company may be cut short.

Third:
Ask your boss to explain to you, "This is how we work *here*."

You can work at a dozen different places, and each will have a different method to their madness of making money. Some will be more successful than others. Find out the process and know what their expectations are. For instance, when I worked for a corporation, I worked on a billion dollar project for two years. Each time my immediate project manager delivered his demands and expectations; I'd raise my hand and ask, "When do you need my numbers done?"

I did this to know exactly how to juggle my time in order to satisfy his needs for the project on my behalf. It occasionally caused work to be continued onwards way into the wee hours of the morning… because I knew he needed his numbers first-thing that day.

The whole point is that you need to know what is expected and how it is expected to be produced so you can best satisfy their hopes and expectations. That requires communication. If you don't know, ask. If you can't deliver on time, make sure your boss knows as much ahead of time as possible so as to alleviate and mitigate any potential problems this may cause.

A tangential quote that illustrates the point-at-hand says, "When it comes to relationships, find someone you can communicate with because sooner or later, everything else goes."

Fourth:
Know that people, even your boss and/or managers, may not know good from bad. They just may align themselves with what succeeded before... because nothing succeeds like success. Said another way, "Success begets success."

Fifth:
Learn to prioritize and set daily goals. Do critical tasks first and get them done before you start checking emails because emails often get you side-tracked and distracted.

When working on a tight deadline, unplug and disconnect from distractions... such as email, cell phone calls, texts, and surfing the internet. Communicate with others how you'd like to be treated while at the office so you don't get overextended and exhaust yourself in the process by helping others do their job.

Have an end-of-the-day ritual that involves scanning your next day's schedule so that your subconscious can percolate over how to make tomorrow as efficient as possible. This also includes cleaning and clearing out the clutter so you don't feel so overwhelmed when you walk in the next morning.

Make sure to take many small breaks through the day. They re-energize and re-focus. Also, sometimes, walking away from a problem lets your mind see it from a different perspective so you come back to it and attack it from perhaps a different direction.

Top 5 Things to Consider at the End of this Phase in the Engineer's Journey:

First:
Did you enjoy the corporate experience? Some like it. Others don't. Which are you?

Second:
Feel good about what you've accomplished and have done.

Then look ahead to see what you can accomplish next.

Third:
Ask yourself if you stepped on a metaphorical landmine somewhere along the way. It's a part of the engineer's journey. It's a part of the artist's journey too. Life is a never-ending learning process. Even if you failed, it just may be Life sending you a sign that your journey really needs to go in another direction. This isn't a bad thing. It's data.

Data is data. It is what it is.
That doesn't mean you have to like it.
It just provides proper perspective to make wiser decisions.

Fourth:
Realize that a lot of people experience a mid-life-crisis due to working a "spirit draining" job. They have tony educational degrees. They make a ton of money. If your heart isn't in it and you're just doing it for the paycheck, re-evaluate if this path is still for you.

I've heard stories of corporate executives breaking down in tears due to not liking their jobs… despite the fact that they make over six-figures a year in salary.

Don't let that be you.

Fifth:
Ergonomics matter. So improve your typing skills. It can alleviate carpal-tunnel issues. Learn to type (and sit) properly for "preventative maintenance." That's because an ounce of prevention is worth a pound of cure. Stretch your fingers and wrists frequently.

Now that you've lived in a large corporation's "corporate culture," ask yourself if you like cubicles or if you want an office where you can close the door.

Know that sound trickles out of "your work area" into others' personal space. The same is the true of the smell of microwave popcorn... which tends to permeate corporate hallways around 3 pm when someone is craving that afternoon snack to keep them going.

Realize that large organizations come with lots of meetings and layers of bureaucracy and administrators who pose a lot of "red tape." Sometimes, if you look like you're having too much fun, you're not treated as trusted.

Lecture Ideas for the Students in Your Classroom:

The importance of "focus" matters. Focus on the "one" thing.

In life, you may feel like you're being pulled in three different directions at once.

Talk about why every answer often inspires three more questions.

Tackle the topic that big corporations tend to hire people who conform to conventional methods and thinking instead of picking up intellectual all-stars and challenging them to think outside of the organization's "box." It's difficult to develop a breakthrough idea in a larger company. Further, a lot of all-stars don't stay long. They tend to leap-frog.

Some Homework Suggestions for this Week:

Have your students consider and describe their perceptions of the horizontal strata that exist in larger corporations versus the vertical structure that permeates smaller companies and then have them include mom-and-pop brick-and-mortar shops in their discussion. One presents a level road with fewer chances of climbing up the corporate ladder. The others are all-risk and all-reward if they turn out to become a lasting success.

My Ruminations about this Stage in my Life:

When I first got hired directly out of graduate school, the company was going through a transition which required a reduction in the workforce... so it was awkward. Furthermore, while they hired a large group of fresh engineers, many of us new-hires didn't have a specific job within a specific section of the company which could utilize our talents and specializations most efficiently. Thus, many of us had to get additional interviews in the division of the corporation where we most hoped to work. This wasn't easy. It was a time of looming layoffs. Bringing in someone new while other experienced engineers were being given pink-slips didn't exactly make me feel welcomed with open arms immediately after being hired. Fortunately, I targeted the antenna department (which was my specialization and forte in grad school), and my hiring manager pulled a few strings to get me re-interviewed within their department. When they accepted me within their division, I felt elated and on-top-of-the-world. On the other hand, be careful what you wish for... you just might get it.

This was true in my case because in order to alleviate the tension of bringing in a new-hire during layoffs, management decided to put me on the graveyard shift so that an "out of sight, out of mind" perception would be achieved. That meant trying to sleep during the day when it's still daylight, when traffic is heavier, and lots of lawn mowers and other power equipment noises made sleep elusive and a fleeting commodity.

That graveyard shift phase of my corporate engineering career lasted for about a year, but the nice thing about it is that it offered a 28% shift-differential increase in my salary which helped me pay off a lot of debts and have a lot of fun-money to spend as a young hotshot engineer living in LA... not a bad trade for the poor sleep I got during that period. The tough part of going back to the day shift was missing the bump in my weekly paycheck due to working nights. Oh well… easy come; easy go.

When I left my corporate engineering position in 1995, I realized, "I wasn't what I wanted to be. I wasn't doing what I wanted to do."

After three years, I felt like I hit the glass ceiling in my corporate career at the company I was at. It felt like having reached the point of diminishing returns, and I was worried about "the glassy-eyed look" I'd seen in the eyes of some engineers who walked the halls in a daze like their heart was gone… lost somewhere during their career. That scared me, and other, young engineers noticed the same thing. I remember a group of us newer hires agreeing on this… with worry and concern in our eyes and tone as we voiced our concerns about how it developed in others. For me, when I departed that career in 1995 to become a writer, a definite "leap of faith" was necessary and required. I can honestly say that the day I made the change, it was the scariest moment of my life.

As I grew to know, most are afraid to try it, but most dream about it daily.

As I learned, the hardest part was the loss of "structure" that required me to learn to fill 24-7-365 somehow, some way. It's harder than it seems. Ask any retiree. It ain't easy.

The thing that makes such a leap of faith difficult is adult debt. This is also what I think leads to that "glassy-eyed look." It's the feeling that you have to work because you've gotten yourself in too deep

One interesting tidbit about this stage of my life is that my "escape" from this daily corporate "Dilbert-like" grind was my quest to play pro-level racquetball… a goal I reached at the end of this stage. Along the way though, there was one opponent who was an attorney who consistently advised me that I'd make a great lawyer and that I should consider entering law school.

Even though I'd already spent a decade pursuing engineering, he'd constantly remind me at our athletic club to consider another career path that he felt I'd be better at... and more financially successful at too.

Then there were other engineers I'd work with who would say, "Brian, you're too creative to be an engineer" as we'd walk from one building to another for a meeting at another location on the corporation's campus. It influenced my departure decision.

I guess they all added up to me eventually feeling like a cog in a very big machine.

On the other hand, it earned me a great living which enabled me to purchase a few items of luxury which I still use today... 20 years later. The lesson I learned is that quality lasts. For example, last night I savored proofreading this textbook in my huge black leather chair that cost me $1100 back in 1995. It's still the nicest leather I've ever touched. Yet, that durability has directly led to my comfort all these years later. I wouldn't have it without having had that engineering job in a corporation.

Another cool thing about my corporate gig is that I was able to earn a lot of overtime when I learned to become the "go-to guy" when it came to designing and analyzing waveguide bends which would minimize signal reflections at the desired frequency range. Somehow, an engineer passed along a formula that calculated the "magic angles" needed when I was inquiring about another bit of data analysis. It was one of those off-handed, "Oh yeah, this might help you too" kind of things where the engineer I was speaking with was going through his stack of stuff. That 10 minute talk earned me thousands of extra dollars once I read the manual and taught myself the 3-D computer aided design (CAD) program called High Frequency Structure Simulator (HFSS).

Finally, the best lesson I learned came from a mandatory training day off-campus. Our managers were required to take this course and appreciated its elegance in shifting the paradigm of accomplishing more efficient proficient procedures and processes. The lesson was taught by a consulting firm, and ten engineers from our division were sent to work together. Those consultants split us up into two teams of five players each. The goal was for each person of each team to sign five index cards five times each. This process was done three different ways. The first way required each person to sign each index card five times before passing it down to the next engineer in the chain. This process was timed. It also was the procedure that took us the longest. The second way was for each team member to sign each index cards three times before passing it down to the next waiting engineer. Then when the last team member signed their name three times, they'd bring that card back up to the first engineer who would sign that card again twice before passing it... in order to achieve the five-signature goal for each person and each of the five index cards. This process took shorter than the prior procedure. The last way they had us complete the task was for each team member to sign each card once before passing it... and once the team member at the end of the table signed their name, they'd bring it back to the first person... again, until each card was signed five times by each of the five team members. This last process was the fastest.

The reason they broke us into teams was to make the process competitive. They may have offered a small prize for the fastest team to complete the task successfully to make it more interesting and appealing. The goal was to show the importance in minimizing downtime of any team member by making minor adjustments in the procedure which accomplished the task successfully.

I realize I mentioned this last chapter, but it's important enough to repeat and understand.

This Week's "Take-With" for the Professor:

Corporations like Ford, General Motors, and Chrysler are a very different experience than working for DTE (Detroit's power company)... Boeing... or GE (General Electric).

However, a good metaphor plays on all levels… just like a good parable.

The whole point is to communicate a general sense of understanding. That way the students can fathom and envision the potential and possibilities which await. This textbook's goal is to provide "an idea" of "what it may be like" at every stage of the engineer's potential journey. Thus, it's similar in concept to "fuzzy math."

Saying it another way, it's like having run a lot of case histories and seeing a statistical trend develop over time.

This Week's "Take-With" for the Student:

Another thing you want to ask yourself is whether you like to travel for work. Are you a road warrior for your job? Do you fly to other states and other countries on the company's dime to pursue new business horizons? How do you like your "per diem"?

Everything counts "One" so take this into account when you decide how much of your weekly salary to invest in 401k plans and stock options that your company matches. The whole point to keep in mind is that if you invest it before you see it, you won't miss it.

Another thing, invest as much as your company is willing to match. It increases your salary beyond what your base salary actually is.

Finally, when it comes to your work ethic, don't make excuses. Deliver results.

Moral of this Week's "Story"

Robustness ensures things work "right" in the real world.

Cutting corners in designs and building the product may save time and money, but they boomerang back on you and your company later on. For example, the Titanic cut corners by not having enough life boats. In the case of automotive companies, recalls have become commonplace… and in some cases, have led to the loss of life.

Important values can't get set aside in the quest to achieve #1.

A good competitor wants their opponent at their best.

<u>Week 10</u>:
Engineering Realities While Working for a Start-Up

Some Brief Brainstorming:

Envision the realities of San Francisco's "Digital Goldrush" and working in a business incubator that utilizes a shared workspace floor plan.

Bring in the best and brightest, but be willing to listen to any of your employees as to how to make your business better. It's an "adapt or die" atmosphere that you'll all face in order to thrive and survive the unexpected challenges and surprises at almost any start-up.

Yet the best and brightest usually only stay less than 3 years because they can find better offers and leap-frog their way up in salary. Complacency and boredom take talent away. Ambitious intellectual renegades typically have a short shelf-life at any company.

Give free stuff to make for a great place to work.

It's the essence of the term "swag" – "stuff we all get."

The more perks you provide; the more likely they'll stay.

It creates a space that fosters more productivity from people.

The Humor in this Stage:

One of life's conundrums is that those who crave power the most… tend to be the ones who are least able to exercise it.

If you're an employee in a start-up usually there will be a "big boss" who is the person who came up with the concept for the start-up to take reality and begin its business. With that person, no matter how casual the atmosphere, a quote from Machiavelli's *"The Prince"* certainly applies, "Don't outshine your master."

Word of warning conveyed. Buyer beware… because once you start to outshine your boss, it's time to become your own boss.

Student Goal(s) for this Week:

Dream of being your own boss to call all your own shots. What widget would you put in the blood, sweat, and tears to rush into the market and swim with the sharks? Realize that if you become your own CEO, you'll probably burn a lot of midnight oil and put in a ton of overtime… which may infringe on the other areas of your life. That balance between work and life is a difficult dance to do for the boss who has so many irons in the fire. It's the prime reason for the slogan, "A leader delegates."

Professor Goal(s) for this Week:

Encourage your students with the entrepreneurial spirit. One way to enable it to flourish in your classroom this week is to create a flow-chart on the chalkboard (or white board) that starts with "Innovation" which goes to "Reinvent old models" then "Package it with the customer in mind" onto "Keep it simple" then end with "Does it save money?"

The Benefits of this Stage in the Engineer's Journey:

One of the nice things here is that if you are the visionary who capitalizes on fulfilling a niche in the marketplace with a widget you do better than the others, fill a need, and beat others to the punch, you can do really well in life.

The Downside of this Stage in the Engineer's Journey:

You might not always get paid on time and with consistent regularity… OR… your demand may outpace what you're able to produce without large capital investments.

The other problem is that you may start to enlist angel investors and/or venture capitalists who try to enforce their agenda on you and the others at top who created the company.

If You Don't Know Your Destination, How Will You Know When You've Arrived?

In a lot of start-ups, they underestimate the amount of time and money it will take to reach self-sustaining profitability. They may not realize all of the rollercoaster ride cash expenditures that can arise in the future.

It's like driving at night and only being able to see as far as the headlights shine as you go down the road. That's where it pays to know where you're going, and if possible, have a map that gives you a lay of the land.

The general average is that it usually takes about 5 years to get all of the kinks out of the system and for your business to achieve a smoother ride. However, in life, "It's always something." So learn to dance in the rain rather than trying to dodge the storms.

This is where you really have to believe in what you are doing so you endure the downticks that will eventually arrive. Sometimes a promised contract doesn't arrive. Other times, a deal falls through (for any number of various reasons) that you needed.

How you handle these downtimes will determine your company's character. That will trickle-down into the reputation you acquire and how people perceive you.

Top 5 Things to Do at the Start of this Phase of Your Life:

First:
Start to develop an elevator pitch in order to promote your start-up's primary product. People will always ask what you do. Having something that immediately intrigues them can and will open potential future opportunities.

The elevator pitch is saying something succinctly so people can get "just the gist" of what you're working on. For example, one PR (pubic relations) person told me that a good product's brand should be able to be presented in five words.

Second:
Try to create a product that shifts a culture. That way, when they walk out from their experience which you made and produced, they have a different perspective. Create something that delivers that impact. Be bold. Create positive change in people's lives.

This is usually never done in one single step. It requires a "gradient of change." It takes many small steps. Usually it also requires "a change agent"… someone who has some experience of that Vision you wish to see manifest. Innovation starts with this Vision.

Deliver a quality product that impacts positive profound change in people's lives.

Third:
Be colorful. Be grey. Be black and white too.

One comment I got in my creativity after working for a large corporation and moving to a start-up was, "Wow, I can't wait to see what you do once you add color."

For example, one engineering professor once asked me how I start writing a book. I replied, "It starts with an idyllic time of brainstorming all the chapter titles… which leads to all the content I want to communicate within the pages of the book. Then I structure them into an order which makes some sort of logical sense… a sensible progression. After, I decide if there's going to be a repeated structure within each chapter. If there is, I brainstorm that structure-within-a-structure next. After, I type all of those into the computer in separate files… that way I can print them and show them to others so they can get 'just the gist' of what I'm writing. After, I create a brand new file which will eventually become the book's completed manuscript file. To that new file, I copy and paste the table of contents with its chapter listings – twice. One will just be the table of contents. The other time I copy those chapters, I add spaces and page breaks in between those individual chapters so when the moment strikes, all I have to do is find the particular place in the file, plant my cursor there, and start typing away. If there's a structure-within-a-structure for each chapter, I copy and paste it into each chapter with white-space between each segment so the manuscript file is writer-friendly. That way, I don't need to write from start to finish. I just start with an organizational outline that is really the manuscript's 'skeleton.' Then I just write whatever is strongest at the top-of-consciousness... at that particular point in the manuscript which fits and is most applicable for serving that section's best-interests."

She marveled over how "structured" my writing approach was and quipped, "Only an engineer would come up with such an organized way to write a book."

I nodded and agreed, "Yet, notice how once I create that backbone, that skeleton, that structure... how much freedom I build into what I eventually write."

She replied, "Only an engineer would think like that."

She reminded me that she then had a meeting with the dean of engineering and all the other engineering department heads so I left her with this, "That's why it's easy for an engineer to turn creative than it is for a creative person to turn into an engineer."

In a start-up, there is a form of creativity you employ... in a structured fashion. Yet as Coco Chanel once said, "Fashion fades. Style remains. The ability to discern between the two never goes out of fashion."

Fourth:
One term I've heard tossed around a lot of marketing meetings is "secret sauce." Really, it's the essence of what makes you or your product special.

The fact is, if your "secret sauce" is baked in truth and reality, you may have a hit on your hands. Should this occur; the sky may be the limit. This is where an IPO may occur too… which means an "initial public offering" of stock. Those are the essence which can make you a millionaire or a billionaire overnight if your "secret sauce" is the next great thing that becomes the new "flavor of the month." Facebook is a prime example.

Fifth:
Consider the following truth about an entrepreneurial start-up: "If you're not ready to be uncomfortable, you're not ready to be an entrepreneur."

Instead of trying to come up with "the next *big* thing;" consider shifting your paradigm to trying to discover "the next _small_ thing" and finding the necessary iterations that keep customers coming back for more after the initial splash of success. The nice thing about this way of thinking is that it requires raising a lot less capital.

Top 5 Things to Consider at the End of this Phase in the Engineer's Journey:

First:
Did your start-up offer creative incentives to improve excellence in the workplace?

If so, how did they affect employee performance?

Second:
Just like in business, it all comes down to "location, location, location."

Was your start-up located at a place which maximized its potential for success?

Third:
Life is risks. With no risk, there is no reward.

One of the biggest risks a start-up can take is to accept an order that is above and beyond their capacity to produce at an economy of scale.

A case in point is a food start-up I worked with which received so many orders from grocery stores for their product that they couldn't keep up without an additional investment from a venture capitalist of $50,000-$100,000 that could automate the packaging process instead of packaging their product by hand. Their problem was that they were too successful too soon so they weren't able to keep up with demand.

They had a great and healthy food product, and they packaged it in a single-serving size which filled a niche in the marketplace which hadn't been addressed yet by any of the competition. This is why grocery stores prominently placed the product of their shelves. The CEO's friends, family, and parents invested as much as they possibly could, but the industry's demand outweighed their ability to produce it at an efficient economy of scale without an additional influx of money to expand their production capacity into profitability; thus, their initial risk turned out to be a flash in the pan.

Fourth:
There's a saying, "Cheap, fast, good… choose any two."

As a colorful side bar, I made up a similar saying for relationships, "Physical, emotional, intellectual… choose any two." Finding someone with all three is rare.

Just as there is no perfect person, it's very hard to make a perfect product. That's why you often have a version 2.0, a 2.1, a 2.2, a version 3.0, and so on.

Innovators iterate.
They compare and contrast.
They see what works best.

Fifth:
Realize that 90% of start-ups fail after your prime angel investor/venture capitalist money disappears. The "bank" (of your angel investor) may close its doors. What then?

What happens when your initial investment money runs out?

The mind will quit five times faster than the body.

Avoiding challenges in a start-up is an oxymoron. There will always be obstacles and ambiguity to tackle. If this style of life isn't for you, you can always consider getting another job in another kind of company. Some people aren't built to work at a start-up. It's okay to acknowledge this fact. Just be honest with your employer. They deserve it.

Lecture Ideas for the Students in Your Classroom:

Accentuate the two roads traveled by most typical start-up companies. One road drives towards going public with an initial-public-offering (IPO) of stock. The other road takes you to the destination of being bought-up and acquired by a large company…usually for a very generous fee which will make the CEO rich. This is where the risk-versus-reward for all the headaches a CEO faces when they launch a start-up is worth it.

Some Homework Suggestions for this Week:

Expand on the quote, "You must pay the cost to be the boss."

Second, ask your students to list their most important aspects of their primary passion.

Third, from that list of their passion's aspects, make them distill those into five words. Have them rearrange those five words into five different orders. Ask them to analyze how each order can convey a different essence and perception of their passion.

Finally, ask them to communicate that passion into a 30-second elevator pitch that they present in front of the class. An elevator pitch is an exercise in public speaking. There, they'll realize that brevity is the great charm of eloquence.

My Ruminations about this Stage in my Life:

In my belief, I've worked for three start-up companies. The first one only had one employee – me – though I often considered my mom to be my secretary at times. The second one was an engineering firm specializing in making optical devices to help analyze and improve sports performance. The third one was for a brand of granola which reached grocery store shelves in the local area.

In my case, my "start-up company" is just another way of saying I turned into a writer who took over 14 years of struggle before publishing my first book. The reason I sometimes considered my mom to be my secretary is that I often lived with her over the years in order to realize my dream into a sense of reality. What I learned is that a lot of people who start their own company underestimate the amount of time and capital it will take to turn into a self-sustaining success. Another thing I learned is that getting my books published and placed on sale in the marketplace provided an air of legitimacy that I didn't have when I was "just writing" for all those unpublished years. Often, my books really amounted to being like business cards which helped me land ancillary consulting and ghostwriting gigs for those who had good ideas but didn't know where to start and hadn't put in the time and practice it takes to write really well. Many times along my writer's journey, I'd remind myself, "Nothing gets done unless I do it."

In the case of the sports engineering start-up, there were four full-time engineers (one of which was the CEO), two part-time technicians (which I was hired to do), and a part-time secretary who doubled as the firm's accountant. They asked if it was okay to view me as a contract employee without benefits who would be issued a 1099 tax form at the end of the year instead of having regular taxes and deductions taken out of my check. Another thing with that company is that they often needed supplies and asked me to pay for them on my credit card... to be reimbursed at a later date. Some of the supplies were electronics components, some were sheets of steel I'd get at a steel-yard, and others were lumber and building supplies that required a trip to the hardware super-store... so it was often an adventure acquiring the needed pieces which they then asked me to assemble and fabricate without any direction of how to do it. What I liked about that was that it really caused me to have to think... really think. Also, while they did eventually reimburse me for the supplies I put on my credit card, there was one time when they asked if they could issue my paycheck late because the money they were expecting from their primary customer wouldn't arrive until we completed a milestone that wasn't finished yet. The look on their faces was priceless when I asked, "Wait, you hired me without any money in the account to pay me?" They confirmed they had indeed done just that, but that they did so knowing the money would eventually arrive with my help... which is why they hired me.

Another unexpected aspect of working for this start-up was that once they demonstrated their prototype product to their major customer, it still had a few bugs in it which I detected only two days prior to their departure to England to demonstrate its use. It caused me to be released. It was known that once they reached their deadline of the product's demonstration I wouldn't be needed, so it was merely coincidental that I also pointed out a flaw in their software which they weren't able to remedy before their departure. What happened is that they asked me to try out the equipment with many case histories to make sure it worked properly. At first it seemed to work fine. Anytime an object crossed the sensor threshold, it would register a "score." That's when I slowed the object's crossing to the velocity of a slow crawl. Error messages resulted. Apparently, the internal computer code had been programmed with a counter that didn't anticipate an extendedly slow threshold crossing. That caused the programming to go haywire, and flabbergasted faces were the result from the rest of the employees. They'd worked so hard to have the programming completed before their departure's deadline, and this one error occurrence caused them two sleepless nights filled with headaches as they unsuccessfully tried to remedy the monkey-wrench I unwittingly threw into the works.

Thus, my release was bittersweet, and instead of having me come in to pick up my last paycheck, they insisted on sending it to me in the mail. In my opinion, I walked away with my head held high because I'd found the one flaw they hadn't accounted for, and it was my belief that it was better that we found it than having the customer find it and come back to us complaining that we weren't thorough enough.

Finally, a few years later, I was offered work with another start-up that got their brand of granola on grocery store shelves via filling the niche of offering single-serving-sizes in a variety of flavors. Most of the time, I handed out free samples of its many flavors during weekend "buy local" in-store promotions... hoping that customers would try, like, and buy our product. Other times, I helped write their website's content and then edited its typos after the website's designer entered all the data. I did a lot of website troubleshooting too. Another thing was being like a taxi at times when a ride was immediately needed.

The problem they encountered was that demand outweighed their production capacities. They packaged their product by hand, and they needed to automate the process in order to achieve consistent profitability.

Thus, regular paydays were (again) exceedingly rare. Eventually, they had to reduce overhead so they could attempt long term sustainability while its CEO looked for angel investors and venture capitalists who didn't ask for too much power and control. When they didn't manifest as anticipated, I was let go. It didn't come as a surprise, but the disappearance of any revenue stream is always somewhat of a disappointment so I can't say I wasn't sad. On the other hand, the product isn't on store shelves anymore either. In the end, it was an education and a learning process which is always a welcome commodity to possess.

This Week's "Take-With" for the Professor:

Just like in your classroom, the leader at a start-up sets the tone for the company's culture and its expectations for the way things should be done.

This Week's "Take-With" for the Student:

People tend to need their own personal space. While synergy works best in groups, it pays not to suffocate anyone.

Moral of this Week's "Story"

There are two typical destinations for any start-up. One is that your company goes public with an initial public offering (IPO) of stock that can pad the pocket book of those who built the company from the ground-floor-up. The second one is that you get bought out by a bigger company and merge under its umbrella of offerings for their customers.

Anytime you have to sub-contract out work, you increase the cost. Thus, if you can do more in-house, your start-up will benefit. So sometimes, it pays to invest in the infrastructure that can accomplish this.

Week 11:
Engineering Realities of Working for a Mid-Sized Engineering Firm

Some Brief Brainstorming:

The Scientific Method rules out the wrong answers to make it easier to arrive at the right one... so it's process of deduction... just like Sherlock Holmes.

The Humor in this Stage:

One scientist described The Scientific Method to a student in front of an audience in a most refined and elegant manner. He then waved his hands and said that this definition actually is wrong. He then quipped that it really means a scientist receives a grant, buys a new piece of equipment, and then gathers as much data with it as they can with the hope that they can make sense of some of it in a way which works enough for them to get another grant once the original grant's money runs out.

Saying it another way, when you buy a new hammer, everything looks like a nail.

Student Goal(s) for this Week:

Consider how many employees it takes for a start-up to eventually be classified as a mid-sized firm... and then how many more it'd take to be labeled as a corporation.

Ask yourself which size firm you think you'd prefer to work at so you can target your job search towards similar sized companies when you graduate and are ready to enter into the working world.

After all, a runner doesn't run a race blindfolded. If they did, why would they "just start running" and expect to arrive at the finish line safely?

Professor Goal(s) for this Week:

Communicate the idea that an engineer can make a profound impact. Just like voting in minor elections, the fewer the people who vote, the more your vote actually counts.

One way consultants create impactful change is through a process of assessment. First, assess your audience. Second, teach what you want to see created or changed in a profound way which makes your customer or clients say, "Yes! I want that." Third, do a second assessment. Fourth, ask if your people changed due to your teaching. Fifth, evaluate if your teaching and education works. Goal: create a transformation.

The Benefits of this Stage in the Engineer's Journey:

The whole point in being an engineer is doing something cool. What motivates you to do it... is the basis of your "religion." In smaller companies, there's enhanced opportunity to ask for exposure opportunities which aren't drowned out by the volume of the number of competing engineers who want the same exposure to cool science.

The Downside of this Stage in the Engineer's Journey:

The harsh reality of working in any job is that they might let you go without a moment's notice... even if you've put in decades. Sometimes companies need to downsize in order to survive. The last one hired is usually the first one to be fired. Other times, they just might feel like you don't fit in within their corporate culture. Face it. It happens.

However, this is where a quick "no" is better than a slow "maybe"... because you may make plans that assume longevity which can go haywire in transient conditions.

If You Don't Know Your Destination, How Will You Know When You've Arrived?

Mid-sized firms offer enhanced opportunities to shine and gain peer recognition; however, they also offer relatively no place to run and hide if you mess up. Thus, be aware that your visibility is enhanced in such a firm. A chain is only as strong as its weakest link. At a poker table, if you don't know who the donkey is, it's you.

Top 5 Things to Do at the Start of this Phase of Your Life:

First:
Data is data. You don't have to like it. Many times, a quick "no" is better than a slow "maybe." Ten mistakes made today are ten things you can remedy tomorrow so you do it better and more correct the next time around.

In science, there is no cheating. You'll take your data wherever you can get it. Keep pushing for more data. That's because, in science, your data is never correct, and it's never complete. Also, science doesn't aim to "prove." It records its findings via statistics, and scientific methods deal with data's extrapolation.

Second:
The hardest part of the scientist or engineer's job is coming up with the correct questions. Also, those correct questions you pose while doing your science must also possess societal relevance. When "the thrill is gone" and "the shine is off the apple," project funding might dry up; thus, adaptability is a necessity. There's always a part of the engineer who has to sell the discoveries you make in order to sustain your science and its quest. For example, there's a joke where a king asks Mr. Faraday, "What's the good of electricity?" Faraday replies, "Someday sir, you'll be able to tax it."

Third:
Remind yourself that this week's theory may be wrong. For a theory, a hypothesis, or an axiom to be correct, it has to satisfy two criteria: 1) it has to be beautiful and 2) it has to work. What's obvious to you may not be obvious to others. You might

be forced to sell the importance of your findings in a way that others understand.

Fourth:
Science and its laws can often oversimplify. Its logic is one which must be learned. Its insight isn't going to come to you all at once. Just as a journey isn't made it one step, it does begin with one step. The importance of that journey is its revelations along the way.

Fifth:
While the past few points are somewhat abstract, this next one isn't so elusive.

Acquaint yourself with where all the office supplies are and raid those cabinets to stock your desk with the necessary tools to do the job you were hired to do with proficiency.

Learn the tricks of how to print your documents to the various printers the company may have. Learn how to run the copy machine. These will be necessities at some point. It's better that you learn them and their locations sooner than later. That's because when you first get hired, your co-workers will expect you to take a while before you start to hit your stride and do the work they hope you produce with professionalism and excellence. It's the learning curve that occurs at any company. That's why a lot of companies don't immediately let your health care benefits kick in. In many cases, they wait 90 days to make sure you stick with the company and adhere to its program.

Top 5 Things to Consider at the End of this Phase in the Engineer's Journey:

First:
Insight is perhaps the most important character trait you can gain and possess. It's a tremendous asset when it comes to sales... such as selling your product, selling the customer on the promise of the product's premise, and selling yourself should you decide to go in another direction. Remember, 60% of a buyer's decision

process comes before talking to sales. This is mentioned because many of you may become a sales engineer which requires a thorough knowledge of the product's workings along with the sales skills necessary to close the deal and earn that generous commission. Furthermore, as I was told by a savvy customer when I sold exercise equipment in undergrad, "If you're in sales of any sort and aren't receiving a commission, you're *in* at the wrong company."

Second:
One of the main goals in science and engineering is truth. There is truth in beauty… just as there is truth in poetry. There is also truth in beyond what can be weighed or measured. For example, you can't measure the love you possess for your family, but you still know it's there… even if and when you disagree.

Thus, this saying applies, "If I hadn't believed it, I wouldn't have seen it." In other words, you have to be looking to the sky at night in order to see a shooting star.

Carl Sagan was a person who would never fake his findings or data. For him, truth was worth more than fame or glory. Truth is the firm foundation you build your reputation on… so that you become an esteemed and trusted source upon whom people can rely.

Third:
There are three axioms that an engineer or a scientist must maintain. The first is that the universe we live in and the world around us is not an illusion. The second is that there are laws to be found which describe this world around us accurately and consistently. The third one is that you must believe the science and engineering you do is worth doing. It helps if you believe that the universe is good and that it's worth your time to do it.

The third axiom is the hardest to prove… especially over a long lifetime. It's the premise of making sure your heart is in it.

While science's question may be, "What's it all about, and how do I fit into it?"… the question of the engineer may turn into, "How do I understand it, streamline it, make it more efficient, and enhance its consumer safety ingredient?"

Fourth:
As with any company offering a product or service, you really want to achieve the perception of having or creating something that makes your client or customer feel assured in knowing "Wow, yes, I want that. I need one. I need that person's genius."

Fifth:
Mid-sized firms are like mini-clans. They have more of a tight-knit culture. Either you fit in or you don't. If you don't, you'll know it sooner than later.

People's natural inclination is to stick with people most like themselves. They can grow afraid of what or who they don't understand. So how did they assimilate you under their umbrella, why did they do that, and then why did they let you go? If you don't ask this of yourself if you part ways with a company, you tend not to pay attention to what made you contrary to their way of doing things when you move on to greener pastures.

Intellectual rebels and mavericks may be welcomed inside the walls of the university, but they are often shunned for their contrarian renegade ways within any company.

Lecture Ideas for the Students in Your Classroom:

Growth has its benefits, but if a company grows beyond the size that created the environment that made it a success in the first place, it might eventually flounder.

Other times, a mid-sized engineering firm begins and originates with an idea to "do it a different way" than all their competitors which enables it to flourish. The problem becomes if they don't enable new innovation to enter once that "new way of doing things" grows obsolete. This is where some entities go belly-up and close their doors.

Some Homework Suggestions for this Week:

Explain why easy answers aren't often elegant (and simple) answers.

Explain why the joy of understanding is a prime motivator in your engineer's quest. (Hint: It is beautiful, consistent, and elegant.)

Explain why the human race has a purpose to add to what is given. When much is given, much will be required.

My Ruminations about this Stage in my Life:

I found it interesting that some of my fellow workers in a mid-sized firm tended to consider me a person who really would rather be a rock star than an engineer. I guess it was all the table-tapping I did along to songs I'd listen to and practice along with as I worked. Sometimes, I'd get a "Hey Ringo, give it a break for a while eh?"

At this mid-sized firm, I was the only single bachelor in a company filled with married workers who lived a more domesticated life. Sometimes I'd share some stories which would make some uncomfortable. So to say I stood out would be an understatement.

The joke I've made about it in hindsight is that I'd come into work in the morning and get into some small talk with fellow workers around the coffee pot in the company's break-room. I'd relate some wild story from a party I attended the prior night... and then afterwards ask them how their evening was. They'd typically answer with a toe-tapping scowl, "Well Brian, I took my son to soccer practice and my daughter to her ballet lesson." In other words, it was an extreme clash of cultures for a never-married bachelor like me... one which led to a short shelf-life at that particular engineering firm.

However, when I interned at a mid-sized engineering firm that developed and specialized in crash-test air-bag sensors, it was different. I was a student, and thus was perceived in a different light. That's because "grown-ups" still perceive students as being young, naïve, trainable, and moldable. They expect more mistakes and faux-pas from them. Also though, they know that you're typically only there for the summer, so their level of toleration is scaled to a different level.

If you come to a mid-sized firm, its face-to-face environment of being able to take a lap around the ring of offices and see every employee in the company fosters an air where if they hire you, they consider it like an "engagement" where they try to discern if you fit in or if you're "not one of us." Every employer has their trial period. It's a fact.

This Week's "Take-With" for the Professor:

Innovation comes from little companies.

Efficiency comes from big companies.

Sometimes the errors cancel themselves out, and you mistakenly still arrive at the correct answer. This is why you have to reinforce the importance of verification redundancy.

This Week's "Take-With" for the Student:

The mid-sized firm offers a balance between innovation and efficiency, but it pays to realize which end it tends to tip towards.

Moral of this Week's "Story"

A mid-sized company is one which borders on growing beyond being a face-to-face organization. The problem with that is that it perpetuates a culture where they stick safely with the people and ideas which they know, understand, and are comfortable with.

Oh yeah, there's one more tidbit I should mention.

Learn how to clear copy machine jams.

Trust me. It happens all too often.

Each copy machine has its quirks.

Learn quick to fix it by yourself.

The others will appreciate it.

Week 12:
Monetary Realities of Each Stage in an Engineer's Life

Some Brief Brainstorming:

Getting a cut versus receiving a straight salary – vested interest

Try to find a job that provides a piece of the pie instead of a straight salary.

The Humor in this Stage:

Former pro football coach, Herm Edwards, said, "You can live like a king for a year, or you can live like a prince forever."

Student Goal(s) for this Week:

Realize that your credit score matters. Keeping it in good health and excellent status enables perks that will affect your bottom line… all throughout your life. As a student, it's important to start to establish credit… so consider getting a credit card in undergrad which should be used only in case of emergency. If you can't afford to purchase an item with cash, don't buy it unless absolutely necessary.

Professor Goal(s) for this Week:

Encourage your students to take an economics course as one of their electives. Their ability to manage their money once they start earning a decent salary can make a vast difference in their overall enjoyment of life and the luxuries it affords.

The Benefits of this Stage in the Engineer's Journey:

Technically, this "stage" is one which lasts all of your life. Money comes, and money goes. You may turn out to strike it rich; however, you can still end up in the poorhouse if you live beyond your means… regardless of how much you make.

The nice thing about engineering is that it does offer a comfortable living with a generous salary (at times… depending on your degree of contribution and level of expertise).

The Downside of this Stage in the Engineer's Journey:

Protocol isn't always the answer.

Don't compromise your standards to "make" something work.

If You Don't Know Your Destination, How Will You Know When You've Arrived?

Life is filled with many different chapters, and in each chapter, it'll seem as if you've turned into a new person.

You might even look back at who you were and say, "That was someone else, not me."

Different motivations for "the pursuit" of your goals present themselves as you age and grow.

Likewise, differing "turbulences" occur too… and then when you have children, your priorities change even more. Sometimes the trouble is discerning the values you care about and can extract so you can distill them into "a maintenance of intensity."

This is where pacing yourself is crucial because burnout does happen, and the candle that burns twice as bright tends to live half as long.

When you're young, it's so easy to want to live fast, die hard, and leave a beautiful body behind. In the grand scheme of things, you realize if you stay on a course like that, you might miss some tremendous moments later on. That's what hindsight can do. It's not about the money. It becomes more about the "moments" you create that make life special with a joy to be alive.

Top 5 Things to Do at the Start of this Phase of Your Life:

First:
If you work for a company that offers matching of your pre-tax investments up to a certain percentage, invest <u>at</u> <u>least</u> that percentage. To an extent, it's free money. Also, maximizing your pre-tax investments helps place you into a lower tax bracket.

The whole point is that if you invest the maximum non-taxable amount at the start of your employment, you'll never miss its absence from your paycheck, and it will work for you over time… where compounding interest grows your nest-egg nicely.

Second:
It's important to realize though that 401K pre-tax investments are for the long term. If you try to dip into them early, you'll get early withdrawal penalties that will eat away at what you've earned. The only reason I'd advise cashing in these 401k investments is if you decide to pursue an entrepreneurial endeavor which needs a large chunk of start-up money. If you do decide to go off on your own and cash this money out, the reason for doing so is to achieve a sense of "escape velocity." This is because any entrepreneurial endeavor usually needs enough money to survive the unexpected ups and downs which you'll need to ride out in order to succeed. Usually, this incubation period lasts around five years before you comb out the knots and start making a solid and consistent profit.

Third:
If you're looking for an investment that does have a degree of non-penalized liquidity (liquid assets are those easily turned to cash), consider investing in U.S. Savings Bonds. While they take 30 years to mature, they do accumulate a small amount of interest yearly, and you can always take a savings bond to the bank and cash it in if you're in need of some quick cash.

Fourth:
Consider setting up a "rainy day" interest-bearing savings account along with your normal checking account. In life, it's

always something. Some unexpected expense is almost always on the horizon… waiting to arrive when you least expect it and are usually least able to afford it. This is where having another account that you don't give much daily thought to can help. It's the essence of not putting all your eggs in one basket. It diversifies your financial portfolio, and it builds your nest egg for future use.

Fifth:
Ask if you're going to work with union employees. The reason for this is that a lot of unionized workers get upset if you do a task "just to get it done quickly" when it's really a union employee's task to do. They'll say, "You're taking a union job away by doing that."

A perfect example was when I was an extra on a film set and decided to help set up folding chairs for all the other extras to sit on while we waited in our "holding pen." The union workers gave me some nasty looks for expediting the chair-unfolding since it was really their job to do. Me? I just had time on my hands and wanted to help in a constructive way to help pass the time while we waited for the film crew to set up.

Another issue with working with union workers is that there may be times when they go on strike. This affects more than just the union employees… so be aware of it.

Top 5 Things to Consider at the End of this Phase in the Engineer's Journey:

First:
Life has numerous moments of acute awareness when a major life change is occurring or about to occur. A good example of this is when you graduate and get your degree. A decision must be made… to continue onwards with your education… or to go out and get a job. One of the realities of this particular example is that if you've taken out student loans, once you stop going to school, you'll be required by law to start paying them back.

Second:

When a piece of the pie is offered to employees, it's my belief that they work a lot harder for the much-desired project's success. An excellent example of this comes from the movie world back in the 1970's when George Lucas was still making the original "*Star Wars*" film. At the time, his product's success faced a staggering amount of uncertainty if he'd be able to pull off his vision for what would eventually become a billion-dollar tentpole franchise. Thus, in order to get his film crew to buy in on producing the best product they could, he offered each of his crew a fractional percentage of the merchandizing profits... such as the eventual *Star Wars* action figures. The result? Each of those people on his crew who received a fraction of a percent of his merchandising profits all became millionaires thanks to believing in what George Lucas believed he could do and what was possible. Those profits also enabled Mr. Lucas to create ILM... Industrial Light & Magic in the very beautiful surroundings of Marin County, California.

The whole point is that at the beginning of the deal, he negotiated his long-term financial success into his contract... when the appearance of profits from an unproven commodity was still uncertain. He negotiated a percentage of the profits... a piece of the box-office pie... and he gave a small fraction of it to everyone who helped him and believed in him.

To me, this kind of vested interest in a company's or a product's success creates a sense of ownership that is sadly absent from the majority of career opportunities you'll probably encounter. If you find such an opportunity though, jump on board. Stick with it. That way, if you help a company win a billion dollars; you get a cut for your efforts.

Third:

Ari Weinzweig, one of the co-owners of Zingerman's Deli Community of Businesses in Ann Arbor, Michigan, hypothesizes that the workplace is very similar to a beekeeper, his hive, and its bees. He proposes that a leader (the beekeeper) can't get inside the hive. That person can't fly to the flowers and produce the honey. What that person can do is maintain the hive for greatest productivity to flourish and thrive. One step in that process for success to result for one and all is to acknowledge the different stages in a bee's life. For example, it first learns to clean the hive. It progresses to making the royal jelly. It moves on to feeding that royal jelly to the queen. Then comes its time when it makes the wax. After that, it defends the nest. Finally, they collect the pollen to produce the honey.

Notice the variety of stages in one single bee's life.

Notice that in our lives, we experience many similar stages.

An example of this is listed in each of the chapters of this book.

Our life as an engineer has many possibilities which we may or may not face. The whole point is to respect each phase in order that we may maximize its potential results. In this world, we must co-exist to survive. Recognizing these phases in ourselves and in others leads to harmonious results for one and all. Employ such insight in your every approach.

(Author's Note: If you'd like to read more detail of Ari's beekeeper hypothesis, it is in his excellent book, *"Zingerman's Guide to Good Leading, Part 2; A Lapsed Anarchist's Approach to Being a Better Leader."*)

Fourth:
Dreaming about who you want to be is in some ways a rejection of who you are.

Fifth:
Pinch your pennies and the dollars will take care of themselves… plain and simple as that… matter of fact.

Have you ever been a penny short of the purchase price? It's a frustrating feeling. You have to hope the store has a take-a-penny/leave-a-penny jar, or you have to plead your case to the store clerk, throw yourself on their mercy, and hope that they don't make you pull out your credit card over something as simple as being one single penny short.

The point is that it is times like those, when you realize how much all those pennies you sometimes overlook or ignore, really do add up and matter.

Lecture Ideas for the Students in Your Classroom:

While many of your students will probably have had numerous jobs, very few of them will have had actual "career" opportunities or positions yet. Thus, you really want to concentrate on preparing them for how to invest their earnings upon their first "real job's" manifestation in their lives. Usually, these financial decisions on how to divvy up their pre-tax earnings takes place immediately before their actual employment begins. In other words, they're hired, but they haven't begun the actual act of "working" yet.

At this time, these financial choices will eventually make a vast impact on their retirement earnings. While it might seem silly to think of such things as a student, it introduces the essence of long-term planning in an arena which will affect them on a personal level throughout their entire life. Preaching economic prudence is the key here.

Some Homework Suggestions for this Week:

Have your students track their every purchase for two days straight. Have them record their every expenditure down-to-the-penny during these two days. It'll provide a complementarity comparison meant to address and improve financial stewardship awareness so their economies-of-scale continue to improve as their futures unfold.

My Ruminations about this Stage in my Life:

The personal wisdom from my past that I want to share in this chapter comes from my former grad school advisor… who was also an astronaut for NASA. He flew aboard one of the space shuttle missions, and he was a vital contributor towards the Apollo program.

Years after I graduated, I requested a series of interviews with him whom I recorded as notes for a space-based love story I wanted to write back in 1996 and 1997. While I decided to set that project aside, his insights are still cherished to this day.

One of the greatest take-withs I gained from those interviews was "a way of thinking." It was structured, logical, and very engineer-like. It came from the question, "What was going through your mind when the space shuttle lifted off and flew into orbit?"

His response wasn't that he enjoyed the ride... though he did... more in hindsight though. That's because during the launch, he was rethinking the if-then set of steps he'd need to act upon should this or that scenario play out on their way to reach a safe orbit. Thus, his mindset was to do "A" until he couldn't do "A" anymore... then to do "B" until "B" wasn't available... then to apply "C" until "C's" steps ran their course... and so on.

It was a lock-step approach that he envisioned as an if-then flowchart in his mind during the course of the ascent.

Relating this back to this chapter's topic of finances, it's always important to have a "Plan B," a "Plan C," and further on past a "Plan D" as well. However, it's best to focus exclusively on your "Plan A" until its options aren't available anymore. It's the essence of having one-foot-in and one-foot-out. That produces distracted actions and scattered focus. It's the difference between a "shotgun" approach versus a "rifle" approach.

If you're experiencing mixed emotions about which approach is right for you, consider evaluating your target. Do you just want to hit it? If so then use a "shotgun." Do you really want to hit the bull's-eye? Then use the "rifle." Again, assess your target.

This Week's "Take-With" for the Professor:

This week's topic is the broadest one in this entire book and its accompanying college course. Thus, don't expect to cover every aspect of financial common sense. Just impart a sound foundation for solid financial choices to be made by your students later in their lives. Remember, an ounce of prevention is worth a pound of cure.

This Week's "Take-With" for the Student:

Remember your youth when you tried various ways to earn some money?

The motivation to maintain at every stage of your financial journey is to keep a youngster's initiative to make a buck.

Just make sure to make that buck – honestly.

The more we cut corners to get ahead; the more potential for catastrophe we introduce into our way of doing and our way of being. Doing things right all the time produces a peace of mind which enables you to sleep at night because you know that you've covered all your bases. While the unknown's sense of uncertainty is an essential ingredient within the mystery of Nature, being able to sleep at night without any potential legal issues clouding your mind is a vital component to looking back on your life someday and knowing that you did things "Right."

Moral of this Week's "Story"

The more centered you are with yourself; the less you let things upset you.

Part of that centering, for me, comes from a story a fellow engineer told me over lunch one day in 1995. He said, "There's a story about a young prince who was taught never to finish all of the food on his plate. The reason for this is that it taught him he never had to have it all. It also taught him self-control over his primal desires as well… a good lesson for one and all, eh?"

Week 13:
The Unemployment Blues. What Next? Considerations

Some Brief Brainstorming:

Make up business cards with your contact information if you
become unemployed. That way, when you attend networking
events, you can travel light and still be prepared. As we age,
more and more people will ask you for your business card. Have
one.

Enhance your job skills via continuing educational opportunities.

Buyouts... don't squander this lump-sum money's arrival.
Make it last.

Fulfilling 24-7-365 after "Structure" is removed isn't easy.

Apply for unemployment benefits. Every little bit helps.

Take an internship if you can afford it. It gets you inside the
company and industry you desire to work within which each
stage your life transitions into and through.

The Humor in this Stage:

You got fired? Congratulations! You now have all the
possibilities in the world available for you to pursue once again.

Student Goal(s) for this Week:

Realize that in order to reach the various peaks in life; we must
cross a valley in between each new mountain we dedicate
ourselves to climb.

Understand that as we age, our priorities start to shift along other
avenues than the original road we started out on.

Professor Goal(s) for this Week:

Reinforce the philosophy of "This too shall pass."

Life doesn't waste any experiences on us.

Talk about what failure looks like. Discuss a time when you failed in life and what you did about it.

The Benefits of this Stage in the Engineer's Journey:

At first, it might seem like you're playing hooky from school… like you're getting away with something because you can simply sleep in for as long as you like… for as many days in a row as you care "not to care" about being back in the rat race. Heck, you may even decide to treat yourself to waking up and getting yourself a beer for breakfast. While it's fun at first, such "fun" decisions can turn into terrible habits. However, it doesn't hurt to engage in some shenanigans and participate in some fun and games during this stage. It pays to set guilt aside from time to time in order to achieve refreshment.

So many people exclaim, "Gee, I wish I could do that."
I reply, "You can. You just don't think that you can."

That's the kind of mindset that helps you cope with all the time you swim within during this unemployed stage of life. If you maintain a healthy sense of "damn the torpedoes, full speed ahead," life will be A-Okay for you, and you might even enjoy it a little.

The Downside of this Stage in the Engineer's Journey:

The air of desperation can become palpable during this stage. At times, you might feel like you're swimming in *so much* time that it feels like you're adrift without any sense of shore in sight. That's where depression can set in. Removal of the typical workweek's sense of structure can be a goal killer. Having a boss who snaps the whip and gives you work with deadlines to fulfill can provide a sense of purpose. Being out of work means that you have to snap that whip on yourself and not lose yourself somewhere along the way. One way this happens is taking on a bad habit or two (or three) that helps you pass the time on a boring day. At times, it can seem like the walls close in so much that "cabin fever" arrives with a sense of withdrawal from world that causes stir craziness.

If You Don't Know Your Destination, How Will You Know When You've Arrived?

No whining. Everything tends to happen for a reason... including losing your job. It might just offer the freedom you need to pursue an unforeseen opportunity which was waiting for your unemployed status to arrive.

The whole point is not to get upset and not to get angry during this phase of unemployment. While it may be a difficult time financially, keeping a sense of good energy and positive spirits acts like a magnet for similar vibrations to enter your life.

Imagine that your life is like a fence. Each time you act out in anger or say something unnecessary and mean, you pound a nail into that fence. With time, repentance and apologies remove those nails. The problem is that those removed nails now leave a hole in a fence which was once whole.

Top 5 Things to Do at the Start of this Phase of Your Life:

First:
Apply for "unemployment." Unemployment insurance financial assistance is available. It requires that you worked within two three-month quarters. January-March is one quarter. April-June is another. July-September is the third quarter, and October-December is the final quarter. If you are eligible, you must apply for at least two jobs a week. You will have to list all of the details for each job you apply for... such as a name (if known), an address (or online address), person contacted and their position within that company (again, if known), method of

contact (such as in person, phone, email, etc.), the job's title, and then the result.

Second:
This period of unemployment can seem like a place in time where all is lost. It's a time where you might experience what is called "the dark night of the soul." However, that's where it's often darkest before the dawn. Often, it's the moment of clarity you need to do some self-assessment as to how you can do things better the next time around.

When career catastrophe occurs, friends and family may bemoan your bad luck, but this is really a time to reply with, "How do you know that this is not a good thing?"

The whole point is that the attitude you present during this down-time can directly relate with the way things play out for you.

Third:
Set a goal to expand your network during this stage where you happen to be rich in time on your hands. Check in with people in your network. Express your interest in being open to entertain new opportunities. If people don't know how to help you, they won't. You must communicate with them "a call to action."

Your career is about relationships… no matter what level or strata you're at. Treat each connection like a precious seed whose soil must remain fertile for future blossoms.

Fourth:
Realize that the power in any negotiation resides in the person who has the ability to walk away from the deal. The ultimate power is in the hands of the creator who can destroy what they create. Often, this enables a later-and-greater, bigger-and-better product to result. This was the case for a website I designed. The first one was a behemoth. It was huge, and it required constant and consistent maintenance. Then I got a job at a mid-sized engineering firm which requested that I take it down or put it into a state of hibernation. I decided to delete it. A year later, I

resurrected a new version of the same URL address, but because I started from a clean slate, I streamlined it into a sense of elegant simplicity with very little need for maintenance. While I was a bit angry about having to take my website down at the time, in hindsight, the more elegant version of its web address wouldn't have appeared or arrived if I hadn't deleted it in order to honor my mid-sized engineering firm's request. After all, they paid me a generous six-figures-a-year salary, so if they needed me to clean up my online reputation in order to present me to their clients for delicate, proprietary work on their behalf, tough choices are required.

Being able to "let go" is a difficult but often necessary necessity for survival.

Fifth:
In the working world, a lot of us spend 50 weeks a year putting knots into our rope, and then we expect to take 2 weeks of vacation to untie them all. Life doesn't work that way.

Take the time to make a list of things you've always wanted to do if you had the time. This is the time to do 'em. This is a great opportunity to dedicate yourself towards crossing a few items off your bucket-list.

Realize that your nest egg of money will eventually run out if you stay unemployed for too long of a period of time. Decide how you want to dedicate that money. Is there something in the back of your mind that you've always wanted to do? Do it.

Think about taking a long road trip. If you have the money to do it, I'd seriously recommend it. The change of scenery might be the perfect remedy you need. Afterwards, you can roll up your sleeves and get back to the job of getting another job and going back to work.

Top 5 Things to Consider at the End of this Phase in the Engineer's Journey:

First:
Ask yourself how many times you set an alarm clock during this stage of unemployment. Being able to sleep in is a nice perk. It's a luxury that often goes under-appreciated.

There's a new term used by doctors to describe people who workout too much… to the point of becoming injured by overdoing their attempts to get healthy. It's called "boomeritis." A body needs its proper amount of rest in order to heal and recover. Many times, we're so addicted to "doing" that we forget about "being." Days of rest promote muscle growth. They also foster a nice breather that produces proper peace of mind.

Second:
Assess what this period of unemployment enabled you to achieve. Take stock in these accomplishments which were enabled thanks to this period of job-less free-time.

Not having a job enables your mind to be less distracted and more focused. These are valuable commodities to possess if you decide to create something from scratch.

Third:
Ask what you learned about yourself due to this period of personal freedom while being without a job. Self-discovery and the ability to regroup and steer your ship into a new direction should be cherished and not perceived as failure.

Think about the mighty phoenix that had to crash-and-burn before it rose from the ashes.

Fourth:
Accepting constructive criticism is a difficult but necessary part of the equation that leads to self-improvement. Learning why you failed in the past helps prevent future failure.

Take the time to critique what went wrong. Was it something out of your control? Was it a trait or personality quirk that rubbed others the wrong way which needs adjustment?

Inquire about the underlying cause for your departure from your last job. If it was your fault, find out why. It may not be easy to hear at first, but once you digest it, you'll learn how best to improve. It offers a needed re-calibration so you can fine-tune yourself into a stellar state of renewed excellence.

Fifth:
Taking a temporary position may not top the list of your job search priorities, but they can lead to an earned opportunity later on... simply because you're on the "inside." There, you can prove your "worth" a whole lot easier than being on the "outside."

Thus, "humility jobs" can offer unforeseen opportunities as well as build character.

Lecture Ideas for the Students in Your Classroom:

One newspaper reported how more younger people now trade happiness over a higher salary. They weigh the compromises they must make with the bang for the buck.

Hoping to "be chosen" is the path of conformity. Compare it with focusing on what you can give and contribute.

Some Homework Suggestions for this Week:

Be acutely aware of how you present yourself... in person... and on the phone.

Consider doing some role playing and then have the students turn in anonymous, honest reviews of how they feel the role-players really did.

Similarly, consider having a class participation event where everyone shakes each other's hand, one at a time. After each handshake, have each person write down on a slip of paper their impressions of the other person's handshake. Then fold those pieces of paper and put them into that person's bag. After everyone has shaken each person's hand, have each student read the slips of paper. Then show them the proper way to shake a hand. Firm grip is important, but it's not a contest. Limp handshakes need to be improved. This is especially the case with females. Finding a female with a great handshake is rare, but it is impressive when it is encountered... so practice it... both male and female alike.

If you'd rather give your students a take-home homework assignment instead of an in-class interaction, consider having them communicate what makes their skill-sets unique.

Ask them to persuade a potential employer, in writing, in less than a half of a page, why they would be pleased to employ such a passionate and unique personality.

My Ruminations about this Stage in my Life:

As someone who left the structure of the working world to become a writer, for me, the genie got let out of the bottle. That genie now knows what life is like living outside of the box... so it's hard going back inside the box again. In a sense, it's like a part of my psyche has turned wild. So going back into a domesticated workplace isn't always easy. That's led to some rumpling of the feathers in workplaces that I've gotten jobs in... which promptly led to me getting let go again when all parties realized it wasn't working.

However, one of the best decisions I made after getting let go from a good-paying job was to, two weeks later, embark on an 8000 mile road trip that revisited every single nook and cranny of my past. I visited every place I used to live... and saw the changes. I visited every single mentor who made a positive impact in my life... and gave gratitude and appreciation to each one. After all, none of them are getting any younger.

It turned into the best trip of my life, and I made sure to ensure I did it "right." I stayed at first class hotels and bed & breakfasts a few times along the way… as a treat for the times I stayed up all night driving to my next destination. After all, a night without sleep is really another day gained, right?

The main point I want to impart here is to not look back at this stage in your life with any sense of regret. The money will run out. Somehow, if you keep your wits about you, more money will arrive. To me, it's like living life as a trapeze act. Each trapeze is the amount of money you have to swing to your next station in life. Once the money is used, the trapeze runs its course, and you will be forced to let go in hopes of the arrival of the next trapeze and its accompanying money that carries you closer to your next destination.

Minimize your "what ifs" in life, and you will learn to live life without any regrets.

This Week's "Take-With" for the Professor:

Unemployment isn't the end. It isn't always a bad thing. Sometimes, it's the clarity that you need to reassess the real direction you know (deep down) that your life should head.

As my undergraduate economics professor often liked to say in reference to stock market investments and making money, "Bulls can win. Bears can win. Pigs… always lose."

This Week's "Take-With" for the Student:

Sometimes in life, you have to be at the helm of your ship and sail it safely into your next port. Other times, you have to climb up into the eagle's nest and ascertain your next heading. Like they say, a leader will climb the tallest tree and say, "We're not even in the right forest." A manager will reply, "Yeah, but we're making progress in the one we're in." There's a big difference between the two. There's a time to lead your life, and there's other times to make darned sure you're making maximum progress.

Moral of this Week's "Story"

When I left my engineering career for over a decade to focus on being a writer and a musician, the senior scientist I worked with for three years who wrote the in-house software I used to analyze and design satellite antennas said the following, "I applaud your decision because you're allowing yourself the opportunity to be more creative in life. For me, I'm happy because I come in each day and write computer code for a software program that the company depends on. I have job security because they need me. I have happiness because I'm creative on a daily basis. They pay me well because of it. You don't have this same sense of security and peace of mind. You're replaceable. I'm not. I'm essential. So I encourage you to find your expertise in a way which you grow into the foremost expert and artist in whatever you decide to do. That's because *the only truly happy people I know are creative on a daily basis.* It doesn't matter what they do… just as long as they add a healthy dose of creativity to each day."

That advice, which I received in July of 1995, was as spot-on to finding the secret of happiness. Enabling a sense of daily creativity makes you feel alive inside. Its dynamic approach brings conscious awareness into the underlying reason you choose each route you take… or pass by. It enables those shining moments that make life worth living.

Find them in your life before it's too late to act on them.

<u>Week 14</u>:
Life as an Adjunct and as a Tenured Professor

Some Brief Brainstorming:

Language matters. So accentuate the positive when you frame any idea for presentation. After all, which would you rather have? A cookie? Or an ooey-gooey fresh-baked chocolate chip cookie? Notice the difference. Use it.

The Humor in this Stage:

Quote from an engineering department chairman to an adjunct: "I can offer you a temporary full-time position."

Now if that isn't an oxymoron, I don't know what is.

That chairman got replaced two months later.

Student Goal(s) for this Week:

Put yourself in the shoes of your professors this week and ask yourself if you'd ever envision yourself becoming a teacher or a professor of some sort.

Professor Goal(s) for this Week:

Speak from experience. Be honest. What do you like about being an engineering professor?

What don't you like about your job?

For me, I loved my students and the freshness they infused into my life. The biggest downside was the commute… which, during the winter, could be grueling.

So communicate your own path to becoming a professor.

The Benefits of this Stage in the Engineer's Journey:

Passing the baton of your knowledge and skill sets to students does matter.

The Downside of this Stage in the Engineer's Journey:

Most adjuncts live below the poverty line… even though they love to teach.

A lot of universities require a minimum amount of teaching experience if you don't have a PhD. This is where being a part-time professor (an adjunct) may be the dues you have to pay in order to be offered a better position later on down the line.

If You Don't Know Your Destination, How Will You Know When You've Arrived?

One way of knowing your destination is to take a look at tenure-track professor job requirements, duties, and responsibilities. They include:

Teach contractual load of classes as assigned by the department.

Assign and submit grades according to your university's schedule.

Keep posted office hours scheduled at hours beneficial to your students.

Serve on faculty search committees, participate in departmental meetings, and engage in activities and pursuit which enable you to stay current in your respective discipline.

Top 5 Things to Do at the Start of this Phase of Your Life:

First:
Realize that teachers create opportunities for young people. Your job is to enable access to content they yearn to learn.

Second:
Build relationships. Encourage engagement.

Remember though that different people learn in different ways.

Third:
Don't become a teacher on a whim. It's not fair to the students.

Fourth:
Determined commitment leads to communicating a lucid education.

Fifth:
Prove yourself and follow through so you get offered full-time status if you're adjunct.

During my first week as an adjunct professor, one of my best friends who was a math professor for forty years told me, "Just keep one week ahead of your students, and you'll be fine." That advice seemed to work really well.

Top 5 Things to Consider at the End of this Phase in the Engineer's Journey:

First:
Did you know what you wanted out of this stage? Did you recognize your finish line when you arrived at it? Were you the needle of the compass which pointed out the correct direction? Did you discuss singular concepts that could be seen as salient points from afar, or did you help your students form an overall gestalt to see the bigger picture?

Second:
Challenge what counts as evidence. A theory which is 99.9% correct still fails every 1000th try. That adds up over millions of cases histories.

Third:

Create impact. Create comparative ideas in the imaginations of your students if you decide to teach another term. Introduce "complementarities" to compare and contrast.

Fourth:
Realize that change which is well-informed is the best you can make.

Fifth:
Adjunct professors are a lot like "hired guns."

They're cheaper for the university because they pay them on a term-by-term contract basis which doesn't include benefits such as health, dental, and vision care.

A PhD matters if you're going to get a tenure-track professor position.

One of the hardest things you'll probably face as a professor is the time when the anonymous reviews of your students arrive. The anonymity allows them to be cruel and vicious if they didn't get what they expected or if the class fell short of their expectations. While the majority of my class reviews were overwhelmingly positive, there were a few haters who ripped me to shreds. That hurt. I had to remind myself that haters hate. Also, I had to see through some of the contradictions a few of them voiced... as there were a few "darned if you do, darned if you don't" reviews. However, when you're highly self-confident, you tend not to pay as much attention to anything that seems contrary. In other words, receiving these reviews remedies the "if it's not broke, don't fix it" attitude.

Lecture Ideas for the Students in Your Classroom:

Ask them (your students) about their favorite teachers and why they remember them as special. Find out the memorable traits shared by these special teachers.

This is what an artist of any pursuit does after they attain mastery; they pass the torch to the next generation and inspire them to love it as they do. Good mentors do that. They pass it on. They share an attitude which motivates others to be the best they can be.

Some Homework Suggestions for this Week:

Have the students in your class do some "professor role playing" by describing a few hypothetical students who are on the border of receiving either a B or a C. Ask them which grade they'd assign to these hypothetical students and why.

For example, some of my hardest grading decisions came from students who were chronically tardy or absent for many weeks. While I gave them the chance to catch up on the work they missed, some of them didn't, leaving me with difficult decisions to make.

Also, you might want to ask how your students feel about classes that were graded "on the curve." This was the case for me in graduate school. One of my classes was so tough, everyone would have flunked it if the professor graded without "the curve" from a straight mathematical view where 90-100% is an A, and 80-90% is a B, and so on.

Then since this is nearing the end of this class, ask them what grade they each think they should receive and why.

My Ruminations about this Stage in my Life:

In my case, I took a 20 year sabbatical from engineering. It was my two degrees – a bachelor's degree and a master's degree – that made all the difference in me getting a job as an adjunct professor. This is where the saying "once an engineer, always an engineer" really does apply. The hardest part though was shaking off the rust. I went back and studied the material a lot in order to teach it. One of my friends, a math professor for the past

40 years, constantly reminded me, "Just keep a week ahead of the students."

He was right. The sooner I prepared; the more confident I became in my presentations.

One of the frustrating things is that, as an adjunct, I wasn't offered an office. This made things a bit difficult when I'd arrive on campus early. With a 40 mile commute, I always padded my time just in case unforeseen traffic issues arose. There were times I'd arrive early and then have time to pass before class… with no office to help in that capacity.

Another issue was teaching evening courses… without an office. Myself, I like starting work earlier in the day so that my evenings are "my time." If you teach classes at night, some creative time management may be needed before you get in gear to go to school.

For me, a lot of my students would ask me about the location of my office. As an adjunct, I would simply point to my backpack and say, "You're lookin' at it."

The nice thing about that was that since most of them carried backpacks too, it provided a bonding moment of levity which most appreciated.

There are two other things I really appreciated as an adjunct professor. The first was my return to the university where I earned my bachelor's degree and also fell in love for the first time. A lot of good memories came flooding back each time I returned to campus. That sense of "a prodigal son's return" wasn't lost on me either. The second thing I truly appreciated during my stint as an adjunct professor is that I was encouraged to write a textbook and to create course material. While none of those encouragements were "official," they were expressed enough times for me to write this book… not a bad thing.

Being that a lot of my "classes" were actually labs, when I received anonymous student feedback after the term was complete, I learned that a lot of them wanted me to walk around more rather than sit at my desk and have them come to me if they had a question. They expressed a desire for more proactive professor interaction. In my defense though, my priority upon starting the lab was to collect their weekly lab reports, grade them, and hand them back before they left so I could offer immediate feedback while the work they just completed was still on-top-of-consciousness. As some students told me privately, a lot of their professors took weeks to return their graded assignments and those who voiced this to me said that they appreciated my quick turn-around-times for grading.

One other "teaching" opportunity presented to me was becoming a docent for a Frank Lloyd Wright home. A docent is a tour guide. Yet, being exposed to his high degree of refinement made my life a better one, and it caused me to study his life and the stories surrounding his genius so I could communicate such trivia and tidbits during my tours. After all, I'm a lot like Albert Finney in the fine Tim Burton film, "*Big Fish*," whose main character *savors the flavor over the facts...* much to his son's chagrin.

Thus, when I was asked about my teaching style, I often responded that I considered myself "more of a color-commentator than a play-by-play kind of guy."

This Week's "Take-With" for the Professor:

Validate the concerns of your students so they feel that they've been heard... also so that they feel that they're getting their money's worth for what they expect to be delivered.

This Week's "Take-With" for the Student:

The value of a college degree is more crucial than you probably realize right now.

Make sure to touch upon and discuss "tenure" and its desired sense of job security.

Moral of this Week's "Story"

One of the toughest things any professor faces their first time around as any kind of professor (adjunct or tenure-track) is the skepticism of the suspicious audience. In other words, it takes a while before your "stage fright" in front of the classroom disappears. Once it does, try to have fun with your students. Make them feel like you're all in it together. The more you make your class enjoyable as you teach them what they need to learn; the more memorable (and recommended) you and your classes will become.

<u>Week 15</u>:
Intellectual Property. Patents.
Inventing for Others. Inventing for Yourself.

Some Brief Brainstorming:

Discuss the intermittent windshield wiper and the movie *"Flash of Genius."*

A memoir isn't a summary. It's a window.

An autobiography is a summary.

A memoir is a snapshot of time.

Both are forms of intellectual property (IP)… in this case, covered by copyright law and registered within The Library of Congress (if you live and create the work in America).

Copyrights, patents, and trademarks all fall under the umbrella of "intellectual property."

The Humor in this Stage:

The term "domestic engineer" is one I encountered recently on the business social media site, LinkedIn. What is it? It's the term a single mother used to describe her work experience. However, when you think about it, children are some of the greatest pieces of "intellectual property" that you'll ever create. Invest in them, and they will be there for you in the future. So upon encountering this term, I had to smile. It was such a unique way for a mother the present job skills that no school really ever prepares us for.

Student Goal(s) for this Week:

Realize that the element of *innovation* within any company is often eliminated or discouraged. This is especially true the larger a company is. So if you have "the next big thing" that is "better than sliced bread," consider taking the time away from your career to engineer its invention yourself so you get the credit and the glory. However, you really have to weigh realistically whether it is worth the price of pursuit. A lot of "flavors of the month" take about five years to achieve profitability. Can you risk that rollercoaster ride and with its accompanying challenges and surprises it presents?

Some people aren't cut out for that sort of life. They'd rather have the safe and sure life where each day is relatively predictable. They'd rather have the steady paycheck.

It's a question to pose to yourself sooner rather than later so you give your mind time to prepare for the window of opportunity when a new, great idea arrives where you must decide if you're willing to catch, capture, and harness its "lightning in a bottle."

Professor Goal(s) for this Week:

Discuss copyrights, patents, and trademarks... copyrights being the easiest and cheapest... which can also be completed yourself quickly (without a lawyer).

Introduce the importance of non-disclosure agreements (NDA).

Cover the importance of paying an attorney to draft complicated legal documents such NDA's and binding contracts which can't be construed as "vague" in a court of law.

The Benefits of this Stage in the Engineer's Journey:

Innovation requires that we try to affect change.

It also needs a cohesive larger vision.

While it's nice to make progress, if it's progress in the wrong direction, it may amount to wasted time. However, the detour you make will always offer insight. It's up to you to make the most of the material you gather… even along the perceived "wrong roads." They just may be the data you need to do it "right" the next time around.

The Downside of this Stage in the Engineer's Journey:

Once you let the cat out of the bag, the person who can get your idea to market first is usually the one to reap that initial wave's big rewards.

Thus, the only person who can keep a secret is yourself.

If you must share your secret, make sure to have an NDA handy… a non-disclosure agreement… which you take the time to have a lawyer draw up in order to avoid ambiguities and/or loopholes.

If You Don't Know Your Destination, How Will You Know When You've Arrived?

Even if you get a patent or a copyright for your work, your job isn't finished. Instead, you then need to switch hats and turn from an inventor into a salesman. Some people just see the legally protected invention as the destination. It isn't. You still have to sell it.

Top 5 Things to Do at the Start of this Phase of Your Life:

First:
As former Apple CEO, Steve Jobs, said, "Start with the customer experience and work your way backwards to/for the technology."

Second:
As Steve Jobs further said, ask yourself (as you create and innovate), "Where can we take the customer?"

Third:

One thing I'd consider is to have a dedicated off-line computer. In other words, it's a computer which never touches or connects to the internet. It's an island to itself.

The reason for this is that intellectual property in the midst of development needs utter secrecy.

If your patent-pending product or process is *that* important, why take chances of getting hacked from the outside world by connecting that computer to the internet?

For example, my drum teacher's recording studio's computer never goes on-line. He's now 73 years old and has decades of music recorded. He keeps those songs safe by having two computers... one for making his music... and one for communicating to the world, distributing his music, and then promoting his music in order to make money.

The point is that you never know how deep a hacker can snoop.

Peace of mind for your intellectual property is vital.

Why take the chance if you don't have to?

Fourth:
Know that mistakes will be made. This is good. It means decisions must be made. The process and the product become better because of it.

Fifth:
When you interview with a company, it pays to ask who will own any intellectual property that you create if you become employed by them. For most corporations, they'll have you sign papers stating that they will own your intellectual property while under their umbrella of employment. That often isn't the case if you get a job with a smaller company or an academic institution. It's a legitimate question to ask at the start so you know later on down the line should you come up with a great idea in the future.

Top 5 Things to Consider at the End of this Phase in the Engineer's Journey:

First:
Many times, solutions, inventions, and innovations are discovered by "filtering down" the wrong answers so you can arrive at the right one. On the other hand, adding more data often increases the chances for "connectivity."

Second:
As an inventor, you'll need to create a theoretical model (or equation) which simulates reality in order to be reliable and useful. If the model doesn't match reality, a part of you has to become like a detective who searches for clues as to why there isn't an accurate match. Often, those elusive clues reside where you least expect them so it pays to take a contrarian approach and mindset every so often. It's like playing devil's advocate to see things from the other side. It's about entertaining the exact opposite of what you expect.

Saying it another way, you may want to work backwards from the desired answer or desired result to see what makes sense. It's the essence of "reverse engineering."

Third:
Try to live your life in "bows"… not in "knots." It offers a more elegant way of travel so you don't cross your path until after you achieve a destination and then, you cross your path coming home to "cinch the bow." Thus, you live a more elegant life. And, you gain more serendipity that way too. Try it. You might be surprised. Plus, it makes your daily path more dynamic by consciously adding bows to your typical daily travel routes.

Fourth:
Know that every pro gets stolen from.

As Pablo Picasso said, "Good artists copy. Great artists steal."

However, that's precisely why you need to protect your own personal intellectual property via copyrights, patents, and trademarks. They legitimatize your creations... your brand... and enable legal rights to protect them earnestly in a court of law.

Just remember, if you do get ripped off, consider it a form of flattery. It means your stuff is good enough. It means you're in the ballpark. It means you're a player (and a person) to be reckoned with.

In Hollywood, a lot of movie ideas begin with the phrase "give me the same... only different." In other words, they take an idea which was a success and spin it with a different twist. For example, "*Die Hard*" on a bus turned into the film "*Speed.*"

Fifth:
There's a saying I came across on a social media website that said, "Being a good writer is 3% talent and 97% not being distracted by the internet."

Lecture Ideas for the Students in Your Classroom:

Try to rent the movie "*Flash of Genius*" and show it this week to your class.

Discuss the power of positive suggestion which is spoken. It puts a power of manifestation into the world. Include the idea of envisioning future goals so that you start to steer your ship towards their arrival. For example, where do your students see themselves in five years... in one year... and in six months?

Similarly, if they have an intellectual property idea, ask them to describe their process of turning it into a reality. While they may not know where to go from the "here" of having the idea, starting the active process of brainstorming different ways to realize its reality needs to be done if that invention or initiative is ever to get off the ground and see the light of day. Have them visit the government's Patent and Trademark Office website.

Some Homework Suggestions for this Week:

Consider the concepts of "repeatability" and "standardized sizes" and then generate two lists... pros and cons... for inventing and developing parts and components that are "standardized" versus those which are "one-and-done."

Imagine the cost of throwing away a car and buying a new one after each time you used it. Then direct your discussion into the directions of "durability," "sustainability," and "planned obsolescence" in relation to commerce and keeping employees paid.

My Ruminations about this Stage in my Life:

When I was a young engineer in my twenties living and working in Los Angeles, I witnessed a lot of wealth around me and came to the conclusion that much of it was made by owning a piece of intellectual property that became a commercial success. Thus, after working for the corporation for three years and attending more than my fair share of retirement parties, I saw what those retiring engineers had acquired and amassed over the course of their life's work by earning an honest engineer's weekly salary with integrity.

It was quite a dichotomy compared to the wealth around me in LA. At my health club, I remember one professional musician talk about gigs where he was paid *"stupid money."* When I asked him what the term meant, he replied, "It's where they pay you so much money it's just plain stupid."

Also, I remember seeing a lot of "patent plaques" on the walls of some of the senior engineers as well. Those are the plaques that are given to the engineer who invents something which the company pays to patent and put into proprietary use since it is the corporation who pays the expensive patent fees and owns the intellectual property. It's in the employment contract you'll sign that any intellectual property you create while you're being paid a salary by the company belongs to that company.

Thus, I decided then and there in 1995 in Los Angeles that the best way to earn a piece of the pie for intellectual property that becomes a success was via the copyright. It made the most economic success, and I'm proud to say that I now own over 40 copyrights for my creative endeavors over the past 20 years as of the writing of this textbook.

However, that doesn't imply that I don't have a patent idea. I do. I sat down with a patent attorney to discuss it too. His advice was to build a working model... a prototype... which proves the validity of my invention's patentable uniqueness. He also advised me to do a patent search as well to test the waters of its existence somewhere in the marketplace. When I detailed how my invention would work, he corrected me and said, "You don't have an invention until you build it and prove it. Right now, all you have is *an idea...* a good idea."

He then said, "Let me share another good idea to prove my point. I have an invention idea I know would make me rich overnight. The idea is to make a pill that you put into a container of water that turns it into gasoline. Do you see what I'm getting at? Sure, it's a good idea, but it's easier said than done. That doesn't mean it shouldn't be pursued though if you have the means to make it into a reality."

What I realized from that meeting was that my invention idea would need for me to earn enough money to make a working prototype. That would probably take a team. That would also mean I'd need to earn the money to pay a lawyer to draft a solid non-disclosure agreement (NDA) which I'd have each person of the team sign before revealing my invention's concept with them in order to protect my proprietary idea.

Along another legal avenue relating to intellectual property, the importance of filing the proper legal claims to your idea entitles you to punitive damages if someone uses your idea without your permission. It will not be the idea itself that is legally protected but the execution of your concept that is covered. That will need to be proved on both qualitative and quantitative grounds. That means establishing a paper trail to the party who uses your idea without permission because it can be very easy to have an identical idea on the opposite side of the world which had no contact with your efforts to establish it as your own. However, if the intellectual theft of your idea turns out to be blatantly similar to yours, having the proper copyright, patent, or trademark will make you and your lawyer very happy that you exercised the due diligence to establish ownership.

This Week's "Take-With" for the Professor:

Knowledge doesn't bring chaos, ignorance does. What you're doing in this week's lectures and assignments is lighting a torch in the cave of "invention." The sooner they start thinking of becoming inventors; the sooner they become the creators of their own destiny. This week's motto should be, "I can, and I will." That "can" and "will" attitude manifests good ideas into reality. Inspire them to create... to find their happiness.

This Week's "Take-With" for the Student:

Imagine your wisdom and understanding as an island. The more you explore it; the bigger your "island" begins to seem. Once you reach its shore, you arrive at the sea of the unknown. Wade into its water and pick up a few sea shells. Bring those sea shells back to land. They are metaphorically similar to the ideas of intellectual property you'll discover and come up with in your lifetime.

There's a saying I'd like to share now. It goes like this.

If a man finds a conch shell on the ocean's edge, could you say that the shell belonged to the man? If the man sold that shell, could you say he owned the money? If the man used that money to date a fine woman and marry her, could you say they were made for each other? If the husband and wife conceived a baby, could you say that the child belongs to the sea?

Moral of this Week's "Story"

Innovation is often like a wave. If you want to ride it, you must swim into its ocean. Each wave only has a small window of opportunity. Your awareness of the tides grows. Either you surf a wave when it starts to crest or you don't. At some point, you get tired of "surfing." That's when you need to get out of its ocean, dry yourself off, and go home.

This metaphorical concept of "going home" after surfing an innovation's "wave" is crucial. That's because its period is destined to end. The problem that seeps in is that people tend to want to stick with their ideas and repeat the actions that precipitated past success. This eventually leads to stagnation. Avoid it by mingling freely in other areas. Allow new horizons to percolate new ideas for ingenious new creative innovations.

Said another way... sometimes you have to find other bodies of water to see if their tides enable you to surf their waves too.

Finally, understand the difference between a copyright, a patent, and a trademark. A copyright via the Library of Congress protects books, art, music, screenplays, visual works, etc... and its legal protection lasts 70 years past the creator's death. A patent protects an invention, and a visit to the United States Patent and Trademark Office's website shows that it is the most expensive of the intellectual property legal protections. A trademark protects a brand, its name, and its unique visual appearance's identity.

Afterword

This section presents one stellar student's answers to the same three questions I asked on the final exam that I gave to all my students in all four classes I taught during my first term as a professor. It was written by a student who works in Detroit's automobile industry as a "Designer." He was one of the most attentive students I've ever taught.

Notice that he writes the final exam questions first – before – answering... a good trait.

1) What do you think your grade should be in this class? Why?

I, personally, think my grade in this course should be an A. I believe that I have shown that I have a firm grasp of the concepts and all materials presented. I also believe that I have shown that I go above and beyond the normal expectations of a student when it comes to reading and assignments, and have received very high marks on all homework and exams. I was even excused from doing the chapter 16 homework, due to being out of town, and I completed it and turned it in although I was not required to do so, along with giving a short presentation of our adventures abroad.

2) If you had excessive tardies, absences, or missed assignments, why should I not grade you down?

If attendance and assignments are part of the grading criteria, then if you do not show up to class and do not have assignments completed, you, by all means, should be marked down. Attendance and work are a requirement of the class. If you fail to complete them, there is no reason you should not be marked down. As you are well aware, this is the way it is in the real world. Your boss does not care why you do not show up for work and complete your assigned tasks. He cares about completing the assigned tasks. If you miss enough work, you will not be working at a particular job for an extended period of time. Preparation for the real world environment should be part of the experience you gain throughout your college career.

The exception to the rule, in my opinion, is if there are extenuating circumstances that are evaluated on a case by case basis. Let's face it. Life happens to all of us from time to time. There are instances where we cannot be everywhere and do everything one hundred percent of the time no matter how much we would like to or how long and well you have been trained through the years. If a student makes an honest effort, above and beyond the circumstances that prevented them from doing the work or attending class, I believe there should be some leeway on the part of the instructor. Obviously and ultimately, it is up to the instructor as to how flexible they will treat the student.

Instructors are pretty smart people. They know when a student is trying to pull one over on them or if the student is telling the truth about their extenuating circumstances. If that is the case, then the instructor should have the option of marking a student down more or less harshly based on the attendance and the quality of the work completed. The only other exception is that if the instructor has days where attendance is not mandatory or the student arranges, in advance, for missed time due to work, trips, vacations, business trips, or other factors. I do think that in this event, more accommodation should be made by the instructors for leniency as was my case just this past week.

In all cases with regards to attendance, tardiness, and homework, I believe it should be at the professors' discretion as to how to treat each student on a case by case basis. There should be some hard and fast ground rules laid out in the beginning of the course but, as mentioned above, there are circumstances which arise and must be dealt with in a timely manner in all of our lives. If the integrity or intentions of the student is not called into question, I think the professor should have a little more compassion and latitude towards that student, if the work is up to par, to grant a bit of a pass to that student. If the integrity and intentions of the student are questionable, I believe that the instructor should reserve the right to assess that student as they see fit. I have had exceptionally rigid professors and very flexible professors. I do not prefer one above the other as there are both detriments and benefits to being instructed by both types. My main consideration is to keep raising my overall G.P.A. every semester to get it as close to a 4.0 as I possibly can before my time as a student is over. That entails playing by the ground rules and policies of whatever professor I might have and insuring that I attain my goal no matter how stringent or liberal they are in their policies.

3) This is my first term as a professor. How did I do? How can I do it better next time around?

OK, did you really think this was going to be short? From me? Really? All of what I am about to write is my personal observances and is nothing against you personally.

How did you do? Hmmm. I think you did a pretty good job overall actually.

Now on to part two of the question at hand. How can I do it better next time? I could tell you were nervous sometimes in front of the class when it first started. I know you got thrown into the fire in a very short time span and were scrambling at first to get things going. The nervousness will be overcome with time. It is just like playing music in front of a crowd. The first time is terrifying. After you have done it a number of times, it becomes very easy and really fun.

A few times, you seemed to stumble over a couple of concepts and were questioning yourself. This occurred mostly in the beginning of the term. That is not a bad thing. I will chalk it up to being thrown into the situation rather abruptly and having a very short time to familiarize yourself with the material and the presentation format. Things did get a lot better in the latter part of the term. I question myself all the time and when I do, sometimes I reveal my greatest potential blunders and have it corrected. That's how you learn and become more efficient in your tasks. Think of it as reducing future entropy. You are already on your way. You have the first and most terrifying step out of the way.

The next time will be easier.

You obviously have a command of mathematics and scientific things. That is also a good thing to have when in the position you are in.

Now that I have that out of the way, I will tell you a few things that have worked best for me and some of my observances of professors over the years. Earlier, I mentioned the ultra-strict professors and the ultra-liberal professors. There are many shades of grey in between as well. The best professors I have had are those that are somewhat more on the strict end of the spectrum. I think you have gotten to know me a little bit over the course of the term. Does it really surprise you? The best professors I had all have a strong presence and always speak with authority; meaning that they know their subjects extensively but are not closed to questioning and review of the material and concepts. By no means am I saying that you are not open to discussion. They also admit when they are incorrect and take all questioning under advisement.

On the state of the curriculum, I think that the book used in this class is a very good text.

They have included a lot of excellent examples and illustrations that are easy to understand. Along with the supplemental power point slides, I think it is a winning combination that helps make the material and concepts a lot easier to understand. With that being said, I would have liked to have seen more "board work" in class. By this, I mean that I would have liked to figure out problems in class as a group as part of the lecture time. Some of the best professors I have had in the past are part mad scientist and entertainer. They know their respective material and are passionate about giving that knowledge to others.

On the assignments, I would have liked to have seen a bit more problem solving. Again, you have gotten to know me a bit; is this really a surprise? The concepts and questions you posed for homework were not without merit though. I do enjoy researching the topics and reporting on it in a conceptual manner as well. We can all do mathematics and we have proven that. It does get a bit boring just plugging and chugging numbers sometimes but that is also something I enjoy. Repetitive action is a sure way to cement the concepts and mechanics of problem solving in your brain. I have always found that a class discussion on a topic or specific problem is a great way to learn. Anyone with questions or a different type of solution shares it with the rest of us and we all end up learning from our triumphs and more so from our mistakes even if it means staying for the entire two hour class period.

Those are just a few suggestions, based on my personal observances, of how you might be able to improve your delivery of the material and concepts for the next class you teach. Again, let me reiterate. You did a pretty good job with what you were handed in the very beginning. I did, in fact, learn a lot. Could it have been more? Sure. Being the curious geek I am.

I will probably finish reading the remainder of the text that was not covered in class. I know, sick isn't it? Sorry, this is just the geek/nerd in me boiling to the surface again. I prefer to think of it as a thirst for knowledge.

Anyway, relax a little more. A little more class work on the board and not 100% from the power point slides. A little more mathematics and problem solving. Have fun with it.

It was nice to meet you and have you as a professor. I have a few more classes to take before I am done. I'm sure our paths will cross in the hallways. Good luck and thank you.

If you are interested in responding to any points above or let me know what you think of this dissertation, please feel free to e-mail me. Keep it brief! LOL!!!

This particular student may be a bit older than me, and he was getting his degree... not to advance his career... but for his own personal satisfaction and sense of achievement. While our *university doesn't allow us to give out an A+ as a grade, if I could have, I would have. He was <u>that</u> thorough in everything he did. When we covered the concept of "entropy," he showed me photos of his closet and his wife's closet as an example of it. His wife's wasn't as clean as his. His was so neat and organized that he actually had his shirts ordered via ROYGBIV – the colors of the spectrum – red, orange, yellow, green, blue, indigo, and violet. It should also be noted that he arrived early to class every single time and was one of those students who made teaching fun.*

I'm honored to have had him in my class.

Final Exam:

(One) Going to school comes at a price. The cost of admission, textbooks, tuition, and room-and-board all add up. Do a cost-benefit analysis of the price it cost each student to take this course and compare it to the value and the return on investment they believe they received for going through it. Is it worth the bang-for-the-buck? Why? Why not?

(Two) How could this course be improved later on for future students?

(Three) Have your students read the "Final Thoughts" section following this page and write an essay on why they believe engineering is the correct path for them to achieve their life goals. Ask them to explain what might detour their path to become an engineer.

Final Thoughts

In poker, there's a saying that applies on all levels of life… not just at the poker table. It goes like this, "It's not the number of hands that you play but the quality of choices you make in the hands you do decide to play that determines your poker tournament success."

In engineering, you must make quality choices; otherwise, catastrophes can occur… some of which can lead to loss of life. Having worked for a mid-sized company that specialized at being expert witness engineers in lawsuits where faulty designs led to accidents which involved human injury (and sometimes death), there is a harsh reality that a poor design put into mass production can cost your company a great deal of money due to litigation disputes. So whatever you do, do it with ethics. Do it with grace.

That's a lesson Frank Lloyd Wright learned first-hand at an early age as an engineering student in Wisconsin. He witnessed a building collapse immediately in front of him as it was being constructed. That catastrophe caused numerous fatalities simply because the contractors were trying to cut corners. The lesson he learned from it that he carried on into his vast and illustrious architecture career is, "You must build real."

Thus, ethics and grace do matter.

They are cornerstones of the person you will be.

That influences the kind of employee you will then become.

At Lawrence Technological University, where I attended as an electrical engineering student from 1986-1990 and then taught in 2015, the provost has a plaque on her wall that states the following (though she doesn't know its author):

Your thoughts become your words.
Your words become your actions.

Your actions turn into your habits.
Your habits become your character.
Your character then turns into your destiny.

Finding our place in the world isn't easy. When things are going too well for too long, it's almost an eventuality that some kind of downturn is on its way. When things look gloomy and dark, there's often no where to go but up. It's darkest before the dawn.

Realizing that your life is off course isn't easy to acknowledge. It means that what you believed was right is now (unfortunately) wrong. It's even harder to do something about it. Staying the course really tends to mean "no surprises." Some people just don't tolerate change well. They don't want to be on that kind of rollercoaster ride. The uncertainty of facing "the void" is frightening. However, it doesn't stop people from dreaming about pursuing their passions and doing what they really love. Sometimes, that doesn't necessary equate with what you're good at though. It means a lot more learning. But look at Pope Francis' first visit to America in 2015. He took on the task of learning a new language, English, as a part of his preparation for the journey… at the age of 78. The result though is that his vast preparation directly led to his confidence.

That's so important I'll say it again. Preparation leads to confidence.

So if you do decide to roll with the changes, keep your moral compass centered and calibrated to True North utilizing ethical decision making and graceful actions.

Also, don't confuse tuning with calibrating. You can tune all you want on an AM radio dial and never find an FM station you really hope to find.

Thus, first calibrate, then fine-tune.

If that calibration makes you realize that you're not on your correct life-course, making the change as soon as possible nips a lot of strife and regrets in the bud. As an old engineering friend from Lawrence Tech reminded me early in 1995 before I made my leap of faith away from a corporate engineering career and into a more creative one as a writer, musician, and artist, "No matter how far down the wrong road you go, turn back."

As I've learned since that leap of faith decision in 1995, changing course is not easy.

In my case, I first faced a lot of envy from people who wished they had the courage to do the same thing. That's when a math professor friend reminded me, "If they don't believe in themselves enough to make such a decision, why do you think they'd believe in you?"

After that stage though, my unproven results created a lot of estrangement from friends and family. Fortunately, on the big day that I literally made my move in July of 1995, a famous drummer friend of mine from my racquetball club asked me to visit him at his home. He told me that he made such a decision when he was a boy. That's when he said that he announced to his friends and family that he wanted to become a professional drummer. Being that he grew up in the ghetto, everyone laughed at him… except his mom. She was the only one who believed. So the day that I was leaving Los Angeles to move back to my hometown in Michigan (so I could save on expenses) that famous drummer told me, "I applaud you for making this leap of faith. I did it myself. But just know that you're going to lose all of your friends and family along the way… up until the point when you 'make it.' That's when they'll come back and say, 'Oh, I knew it all along.' After that, they'll all want something from you because your status of having 'made it' will offer perks they'll want. It's a fact that it will happen. The whole point is 'just don't give up.' It's only a conductor at the helm who can keep a train from coming down the tracks. If you stop though, you'll never make it to your destination."

That's when I asked him what the destination was like for him.

He gazed out wistfully, smiled, and replied, "Paradise."

That sense of "Paradise" isn't really an actual physical destination though. It's a state of mind. It's how you do things in your own special way. That's really the goal: to achieve "your own thing." That's where someone can discern it's you without even having to be told it's your work or your creation. For my drummer friend, it's that he achieved his own signature sound. For Frank Lloyd Wright, his style is so obvious that "even a blind man can see it with his cane." For Picasso, the same is true of his art.

In a similar vein, the senior scientist I worked with daily told me something that really resonated when I announced I was departing my corporate engineering career. At first he was concerned that his actions may have precipitated my decision to depart. I assured him that his daily mentorship towards my engineering excellence which helped our company win a billion dollar satellite contract was the one thing I was going to miss the most. That's because he guided me at work like a father would do for his son. His last words of wisdom before my departure were as follows. He said, "To me, you're making the right choice because in your particular job here at this corporation, you don't have a lot of creativity within your job. This decision you're making will provide you with the creativity a person really needs to be truly happy on a daily basis. That's because the happiest people I know are creative – daily. It doesn't matter what kind of creativity they employ. Just as long as the way they do things enables their creativity to flourish. When that occurs, I've noticed that they're consistently happy. That's hard to achieve in life, and honestly, I feel it's one reason a lot of people go through a mid-life crisis. In my job, I have a lot of creativity because I write the computer code that our company has its engineers utilize. Since I write this code daily, I am creative daily; thus, I am happy daily. That's the reason I've turned down offers to go into management. While I might make more money doing so, I know deep down inside that I wouldn't be as creative as I am now. In the job you're now leaving, you didn't have the same kind of creativity I have simply because you just run the computer code that I write. Sure, there are moments of creativity for you. I realize that. But I think you realize that it's too far and few in between. This is why I feel that you're making the right decision by taking this leap of faith."

After those words of wisdom and some pep talk... about how he believed I'd do well because he believed in the quality of person I was... I didn't see him for 18 years. That's when I went out of my way to make an 8000 mile road trip to express my gratitude to all the mentors who made a difference in my life. It was the kind of journey that was a huge full circle for me, and it was the best of my life... despite the grind of making that journey in 20 days... averaging 400 miles a day... not easy at all... but *so* worth it.

When I knocked on his front door unannounced 18 years later in February of 2014 and sat at his kitchen table and told him my story of becoming a published author of over 30 books, he said the one thing that made that grueling 8000 mile trip worth it. He said, "Wow, you left a boy and came back a man."

That comment right there made it all worth it... every single bit.

It's similar to the "Paradise" that the famous drummer told me I'd eventually arrive at if I didn't give up. And yes, I did visit that drummer during that trip too. For him though, it wasn't what he *said* to me 18 years later. It was *the look* he shared with me when he saw me standing in his driveway as he came home from teaching class at USC. You see, he also taught in Nashville at a traveling clinic for musicians called "The Sessions" and was wearing a baseball cap with "The Sessions" embroidered on it as he pulled into his driveway that morning. Being that I had attended "The Sessions" clinic when they taught it in Detroit, I had the same hat, and fortunately, it was packed in my trunk. So when he pulled up and I saw him wearing his, I immediately popped my trunk, put on mine, turned to him and smiled. His nod of approval? Priceless.

We hadn't even said a word, but that look… that expression… spoke volumes.

It was a passing of the torch sort of look.

Those are hard to achieve.

For me, upon each gratitude departure from the many mentors I visited along that 8000 mile road trip, there was an inner-fist-pump moment which occurred immediately after waving goodbye. Sometimes, tears would arrive. Tears that I did it. Tears that I expressed it. Tears that I told the people who made a difference in my life that their contribution years prior really mattered. As I learned after those 8000 miles of gratitude, not enough people do that these days… express the gratitude and appreciation for the contributions people make in our lives. First comes appreciation. Then comes the gratitude. You have to have an "a-ha moment" of realization that those former mentors were really right. That means that a lot of maturity must occur first though. That means a lot of changes and a lot of growing pains had to be tolerated and endured.

A rose must grow through a lot of dirt to bloom. An unexpected thing that 8000 mile road trip showed was that a lot of the places which were daily parts of my past have now closed and are gone. It showed me there's no going back. Knowing that motivates me forward. It causes bridges to be built towards future destinations as you cross them.

As a side note, if you care to read about that journey, its title is *"Gratitude Miles."*

A lot of people talk about changing. Few ever do it. The reason is usually insecurity, lack of self-esteem, and/or low self-confidence. Also though, it's the perception of weakness a lot of people know they'll encounter if they decide to steer away from being on "the fast track" in a respected career.

In following my own bliss, you can't imagine how many times I was told "it's time to draw a line in the sand." You can't imagine the number of times people said, "Maybe you should think about a Plan B Brian." The problem is that the moment you start steering your ship towards a "Plan B kind of life," your "Plan A dream" starts to die. That leads to an inner stagnation which will ring hollow inside your soul for years to come. Being stupid enough not to know when to give up *can* be bliss.

Really, it comes down to how badly you want to reach your promised land. Are you willing to continue to persevere despite the surprises you're going to have to face? In life, it's always something. Can you continually innovate?

If you can, congratulations! You are in rare air. Just make sure to care to share.

If you do, consider yourself amongst the few true "entrepreneurs" in this world.

Some may even say that you've become a "Renaissance Man."

However, that's usually only said after you "make it."

Before you "make it," you may get called "a bum."

Sadly, I know. I was called that by my first love.

Irony's sting is that she used to call me "Prince."

It was two years into my leap of faith in 1997 when financial difficulties caused me to call and ask her for a loan. She refused. I asked why. I then said, "What? Do you think I've gone off and become some kind of bum?"

She replied, "Well, yeah!"

I immediately hung up.

Boy, that one hurt.

It's the kind of bullet you can't dodge.

It's the kind of bullet they never tell you about at the beginning of your leap of faith when you know deep down inside that you're going to face a lot of surprises. There are so many compromises you'll face. There are so many cross-roads you'll arrive at.

In my case, I got offers to go back to the same engineering career I once had. It was the same job… just at a different company. The reason I turned them down is that *I hadn't achieved what I had set out to do yet.* You must pay a cost to be your own boss.

The funny thing is that hindsight has often showed that sticking to my guns and continuing to believe in my dreams has been the correct path after all… even when I had no map on which direction to take, which choice to make, and had to make it up as I went. For example, one of those well-paying job offers I received and turned down turned out to be a company that went out of business years later.

Those are the sort of occurrences that happen along the way that I like to call "Cosmic Winks." It's as if Life sends you a message that only you get. When those arrive, they're special.

It's like being Bugs Bunny and having God throw you down a carrot or two.

If you try to share it with another, they'll never get it. Some may even look at you funny. That's because they'd need all the years of context you alone possess to truly "get it." Yet it's those acute "moments of awareness" that make living Life really worth it.

Legendary Green Bay Packers football coach, Vince Lombardi, talked about this in his *"What it takes to be Number One"* saying. There, he says, *"And in truth, I've never known a man worth his salt who in the long run, deep down in his heart, didn't appreciate the grind, the discipline."*

Becoming an engineer takes discipline.

It isn't easy either. It's a grind.

Honestly, I'm really glad I became an engineer. No regrets.

Though it might not seem like it from some of what I've shared here, becoming an engineer was crucial. Over these past 20 years since my leap of faith, it has enabled me to see the forest through the trees. It's an education that never disappears. The saying that applies is "once an engineer always an engineer."

For instance, this year in 2015, I came back to teach engineering at the same school I got my bachelor's degree in 1990... at Lawrence Technological University. Teaching there earlier this year was an elegant dovetail. Also nice is that the job landed in my lap. I was visiting campus to have lunch with an old math professor friend on the same day the first copies of my latest published book had arrived. Being that I was promoting the book to certain people in town who could increase its exposure and its success, I had extra copies of my resume with me which I was handing out along with the new book. When I arrived on campus, I was early. I decided to stroll the halls and reminisce a bit in order to pass the time. Doing so got me into a conversation with a secretary for the Assistant Dean of Engineering who overheard our conversation, remembered having me for his thermodynamics course, and then asked me if I had ever considered teaching. Classes had begun the prior week, and they still had courses they needed to fill with a professor. So he told me the two chairmen with whom to drop off my resume if I had them handy. Since they'd been sitting on my front seat due to having them out to promote my book with radio stations in the same vicinity as Lawrence Tech, he urged me to do so immediately... sooner rather than later... despite the fact that my lunch appointment's time was now about to arrive. My old math professor understood. He waited while I went and distributed my resumes to the appropriate chairmen. The next day I received a call from both and was asked if I could start that night. It happened that fast.

That, to me, is serendipity.

Luck is when preparation meets opportunity.

So I guess I was just lucky that engineering teaches this kind of preparedness.

As I was told in Beverly Hills while I was acting as a cameraman for a lecture series, "When you are ready to manifest your destiny, it will come to you."

So having this opportunity to teach engineering manifest like it did? It did seem destined. Also, it came exactly when I needed a financial break to arrive. A drastic change had occurred the prior month in my household's finances and another source of income was needed to replace the one that had just disappeared. The elegance of it is that I wasn't even on campus trying to get a job. I was just there to have lunch with an old friend. But by arriving early and being prepared, I got lucky when the needed opportunity presented itself by being at the right place at the right time. The bonus is that it became a nice homecoming too... teaching at your old school is too cool.

As the term progressed, students and fellow professors found out that I'd published a lot of books. Many of them told me I should write a textbook.

When the term ended and I wasn't offered any summer courses to teach, time allowed the idea of writing an engineering textbook to percolate. After August arrived, two weeks prior to the start of fall classes, I had the idea of what to write. Instead of writing a textbook on a specific aspect or discipline of engineering, I'd just write about its journey.

It's one of those things they really don't teach you in school. They should.

One thing that engineering *does* teach though is a mindset. It teaches you how to approach a problem. It teaches you a logic and an approach to tackling a tough task. It teaches you a mentality of extreme thoroughness to ensure that every "I" gets dotted and every "T" gets crossed. That's why I feel fortunate to have earned an engineer's education and experiences before making a leap of faith to embrace my creativity. There's a certain tenacity that comes with being an engineer. After all, the scientific method is about ruling out the wrong answers so you can eventually deduce the right one.

When I left my corporate engineering career, I remember a part of me naively thinking that I'd probably never return to it. But if there's one thing that life taught me, it's to never say never. You never know what Life's tides will bring.

Another thing though that corporate engineering taught me is that 100% focus is what is needed to win billion dollar contracts. I learned that in 1995 just prior to my departure's leap of faith. As that project's leader confided in me before I left, he expressed that the real reason we beat out all the other companies to win a contract of that magnitude was that we focused more than any of the others. We prepared twice as much material as any other company too. That led to the project manager's confidence when he pitched the customer with our winning proposal. Every detail of its design was covered in advance.

I remember being a bit incredulous after we won it. It gave me the courage to walk into my manager's office to ask for a 10-15% raise. Fortunately, its announcement coincided with the timing of our annual raise review meetings. These were done privately on a one-to-one closed-door basis. The incredulity came when he told me how much of a raise I was going to receive after telling me that my government security clearance had finally gotten approved (which had cost the company tens of thousands of dollars). He offered me a measly 2-3% raise with a $100 spot award, and to be honest, I felt insulted.

I had given the company my days, nights, and weekends whenever they asked… in order to win that contract. It took two years to win. Since I was the one who generated the numbers for the antenna's performance, my job was considered a "single point failure" kind of position. That's because the satellite system's team of 60-70 engineers all had to wait for me to generate my numbers before they knew how much power they had to work with. The longer it took me, the longer 60-70 employees waited around. So stress was a constant companion of mine for those two years. With a billion dollars on the line, an unspoken expectation was that I set my personal life aside in order to win the contract.

One time, I got chewed out for leaving my pager at home on a Sunday when I went out on a date with my girlfriend. Another time, I got my butt handed to me for playing in a racquetball tournament all weekend. It was special because it was hosted by my own club… so all my racquetball buddies expected me to be there. My boss was furious.

I was trained to ask "how high?" when I was asked to jump.

When they needed something done overnight, I worked all night.

That's why I felt like I'd been stabbed in the back when I was told what my raise would be during the annual raise review. After all my grueling efforts, it felt like I'd gotten screwed. The billion dollar project's leader concurred. He told me so later on in private.

So when my department's manager told me what my raise would be, I replied by saying that if that was all he was going to offer, I was going to resign. He shrugged and said, "That's all there is in the budget." I reminded him of all the money the company had just spent getting me a security clearance and that that money would go down the drain if he couldn't give me the raise I felt I deserved. He reiterated, "It's just not in our budget to give you a larger raise." I then said something like, "You've got to be kidding me. I just helped win the company's largest contract in its history of existence, and you can't give me anything more than 2-3% and a hundred dollar spot award?!?"

He didn't budge… so days later; I turned in my two week notice.

And by the way, I didn't get awarded the hundred bucks either.

During those two weeks, a lot of my fellow antenna engineers asked me why I was leaving. I was vocal about the reasons. Apparently, over the next year after that, a mini-exodus of talented engineers occurred. So two years later, when I was back in town and having dinner with a few of my old engineering buddies, they told me that they all had just gotten 10-15% raises in order to retain their talent since so many others had jumped ship and went to competing engineering firms. They all believed this was due to me.

They said it was my stand to leave the company instead of staying and accepting their measly raise that precipitated a change in the company's policy.

In hindsight, I guess it's similar to why unionized workers sometimes go on strike.

In my opinion, I remember feeling like if I accepted their raise; a part of me would die that day. It was the moment I knew my life was off course and something drastic needed to be done. For years after I left, my mom would ask if I'd gotten fired and was covering it up just to save face. It got to be a point of me rolling my eyes each time she'd re-ask.

One thing that aided my decision was that I was 27 years old, single, had no kids, and didn't have any debt. It was the only time in my life I was completely debt-free. It made the decision easier. I knew a window of opportunity had arrived. I climbed through.

The hardest part was that my first book didn't get published until the end of 2009. As you can imagine it was a large gap to fill before a door finally opened. I lost a lot of friends along the way because of it. I lost a lot of family ties too. They stopped inviting me to holiday events. It was just as my famous drumming friend predicted back in 1995.

But that book's publication is when a different point of view presented itself. It led to local TV interviews. It led to me giving talks. It led me to helping others get published. It led to getting a few paid ghostwriting gigs. It led to royalties that helped me pay the bills. However, the one thing I didn't expect was that it would lead to a fresh perspective.

That new way of looking at things arrived when I was giving a talk to my church's men's group. I spoke about my journey away from engineering to become a published author. I mentioned the reasons why I left my corporate engineering career. That's when one man raised his hand and said, "It seems to me as if you finally found your place in the world. Instead of being bitter, maybe you should see it as a blessing. Imagine what would have happened if you'd gotten that raise."

He was right. As I look back, I don't have any regrets. I haven't lived a life of silent desperation. I have traveled what I feel has been my destined path.

Looking back at another aspect of the journey, a lot of people have agreed that it's probably a lot easier to go from being an engineer to turn into a writer, an artist, and a musician than it is the other way around. It's a lot harder for a creative person to become logical than it is for an engineer to turn abstract. The discipline is different. Also, once you get out of the school-going mode, it's harder to come back later on after some time away. Rust tends to seep in. Going to grad school immediately after undergrad really helps. You haven't forgotten anything yet. For me, grad school taught me I can teach myself anything I set my mind on learning.

That grad school mentality taught me how to learn a lot of new tricks… such as how to write well, the publishing process, performing music at a world-class level, painting huge public murals, and finding my own unique Voice along the way.

Yet it's the engineering mentality that helped me keep my nose to the grindstone during all those years when no one believed. You just can't imagine how many people thought I'd gone crazy for leaving a good career. It even caused me to move to a new city, Seattle, for 15 months from 1997-1998. I did so because there, no one told me I was crazy. They just heard that I was chasing my dream and therefore, encouraged me. In Seattle, no one told me "You can't." They only told me, "You can." That kind of paradigm really mattered because it happened at the stage in my journey when I most needed to achieve "escape velocity" to break free from the gravity of past perceptions.

You just can't imagine how many people kept trying to pull me back to my old life.

20 years later, the one thing I've learned is "Once an engineer… always an engineer."

It's why I was able to come back after 20 years away and teach engineering at the university level. When I told the school's provost (who was my former physics professor) about becoming an adjunct professor, she admitted that if she were to go back and teach physics that she'd probably have to become like a student again in order to relearn all the math she'd forgotten over all her years as an administrator.

I nodded, smiled, and told her about my former math professor's advice to "just keep a week ahead of the students" with my studying and class lecture preparations.

She understood perfectly and expressed that she was glad to see me "back home."

While I was offered classes to teach in the fall of 2015 as I sit and write this, I decided to turn them down in order to focus 100% on finishing this book. Some say it's like I'm taking a sabbatical. The head of the school's architecture engineering department applauded the decision by saying, "If you were teaching this term, your mind would be going in three different directions at once, and the textbook would never get done."

I agree. As winning that billion dollar contract in 1995 taught, "With focus, anything is achievable." And who knows what will happen when this book is complete and published. Something may come along that hasn't arrived yet to present itself.

In other words, you never know what the tide will bring.

While life hasn't always been the easiest financially for me over the past 20 years, it has provided a wealth of experiences, understanding, know-how, and wisdom. It's brought me to live through almost every stratum that engineering has to offer. That's why I wrote this book the way I have. It tries to teach the journey. I hope you've found it enjoyable.

Each one of you will have to decide if this journey is for you... should you decide to accept its mission. For me, I don't regret it a bit. Once an engineer, always an engineer.

For you Trekkies, "Live long and prosper... Engage."